Over 1,000 Places to see the Total Solar Eclipse August 21, 2017

City, State & National Parks, Campgrounds & Attractions, Road Trip Planning

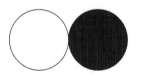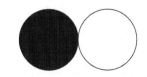

Craig Shields

Over 1,000 Places to see the Total Solar Eclipse August 21, 2017
City, State & National Parks, Campgrounds & Attractions, Road Trip Planning

A Clock Press Book copyright © 2016
by Craig Shields
published by Clock Press

ISBN: 978-0-9846718-4-7   craigshields@clockpress.com

Eclipse predictions courtesy of Fred Espenak,
NASA/Goddard Space Flight Center, from eclipse.gsfc.nasa.gov

Maps courtesy of the U.S. Geological Survey and NationalAtlas.gov

Published in the United States by Clock Press, Kansas City, Kansas.
www.clockpress.com

Disclaimer

Locations listed in this book, such as eclipse centerline highway crossings, are listed as reference points only and may not be suitable or safe for any activity. Some locations may have few, if any, services such as fuel, shelter, restrooms or food. All parks, roads and attractions are subject to hazardous conditions, lack of maintenance, scheduling problems, weather, and other adverse situations. It is up to the reader to select an eclipse viewing site that is appropriate to their needs and safe for their planned activities. The author assumes no responsibility for the dangers, risks and liability that may result from the use of the information in this book.

It is not the purpose of this guide to reprint all the information that is otherwise available to amateur astronomers, eclipse chasers and the general public, but to complement, amplify and supplement other sources. This book does not attempt to cover all possible eclipse watching activities and events.

Every effort has been made to make this book as complete and as accurate as possible. However, there may be mistakes both typographical and in content. Therefore, this text should be used only as a general guide and in combination with other texts, atlases, maps and official information from property owners and governing authorities regarding eclipse watching activities.

The purpose of this manual is to educate and entertain. The author and publisher shall have neither liability nor responsibility to any person or entity with respect to any loss or damage caused or alleged to be caused directly or indirectly by the information contained in this book.

Finally, the reader is encouraged to use common sense, obey all traffic laws, exercise patience and practice safe methods of viewing the total solar eclipse on August 21, 2017.

Index

Every State Section Contains:
- Overview Map
- Eclipse Track Notes
- Larger Cities and Highways in Totality
- Weather and Eclipse Related Links
- Featured Eclipse Destinations
- Centerline Highway Crossings Table
- Eclipse Location Listings - West to East

About this Book

Eclipse locations are organized by state from west to east. Each state section includes a map, eclipse track information, cities and highways near the centerline, weather related links, featured destinations, an eclipse centerline highway crossings table and the eclipse location listings.

CP in italics for a location indicates that this listing is an Eclipse Centerline Highway Crossing Point. **The centerline is where the eclipse lasts the longest.** More information on crossing points is found in the Eclipse Centerline Crossing Points table in each state section.

To see a map for any location in this book:
- Scan the QR code with your smartphone and a QR code scanning app to launch the map in your phone.
  or
- Visit www.clockpress.com/eclipse to see online maps for all locations listed in this book.
  or
- Type the GPS co-ordinates into any online map site.
  or
- Search for the park or site name in your favorite online mapping site. Google Maps works best.

Totality start time and duration are shown for every location in this book. All times are local to that location. ***Plan to be at your eclipse viewing site at least 2 hours before totality starts.***

About QR Codes

QR codes are used in this book to store GPS coordinates and URLs for websites and videos. These hyperlinks on paper can be read by a smartphone using a QR code scanning app that you load onto your smartphone. There are many QR code scanning apps available. Bakodo is the authors choice for a reliable QR code scanning app as of this writing. **Using QR codes is optional, see below.**

Scanning QR Codes

If you hold the phone up over the printed page the QR scanning app will scan the first code it sees instead of the one you want, so there is a little trick to avoid scanning the wrong code:
1. Launch the app.
2. To isolate the code you want, touch the camera of your phone to the code you want to scan.
3. Lift up slowly until the app scans the code.
4. The scan will launch the mapping application or web browser in your smartphone depending on the link type.

If you are not using QR codes

Type the GPS co-ordinates for a location into any online mapping site to go to that location on your computer. If a URL is shown, type that into your web browser or simply do an internet search for the place mentioned.

Anatomy of an Eclipse Location

QR code. Scan to launch the mapping application in your smartphone. ➡

Location Number

7 Gleneden Beach State Park ⬅ **Park or site name in bold**
Gleneden Beach, OR 97388
44.87648 -124.03693 ⬅ **GPS co-ordinates**
Totality Starts at 10:16 / Lasts 1:58

➡ **Totality start time and duration**

CP = Crossing Point

8 CP 3420 US-101
Depoe Bay, OR 97341
44.84419 -124.04730
Totality Starts at 10:15 / Lasts 1:59

A site listed in italics means that this is an Eclipse Centerline Highway Crossing Point. See also the Crossing Points Table for the state.

Total Eclipse of the Sun - Monday, August 21, 2017

Eclipse Viewing Site Selection

Choose a Location

Consider one viewing site as your primary destination and have a couple of backup sites in mind, east and west of your primary site, in case of clouds. Watch local weather forecasts and check satellite images for any storm systems in the region just prior to the eclipse. Check with the DOT for road closures and construction in the region and plan ahead for delays. Have your trip preparations done in advance and leave early. **Plan to be at your destination at least two hours before totality starts.**

Viewing Site Factors:

- The farther you can see the better. A hilltop, lake shore or beach will have a lower horizon.
- Trees for shade and wide open areas.
- Flat grassy areas for tripods and lawn chairs.
- No big lights or signs. More on lights below.
- A clear view of the sun at eclipse time.
- Do you want to attend an organized eclipse viewing event or make your own party?

Decide on the factors that are most important to you and keep weather in mind when making your final selection.

Visit your primary site before eclipse day if possible. Be there at the same time of day that the eclipse occurs to check that you have a clear view of the sun. Look for any big lights on poles or buildings and plan to position yourself to face away from them. Consider choosing a different place if there are many lights because they might come on during totality and spoil your view.

Light Pollution

At twilight, street lights, parking lots and signs light up. Cameras flash. During the totality phase of the eclipse these lights can cause a major distraction for eclipse watchers and photographers.

Street lights and signs near the largest organized eclipse watching events may have been adjusted to remain off during the eclipse. But few cities in the path of totality can afford the time and expense to disable and then reset all the photoelectric switches controlling the street lights just for this eclipse.

How to deal with light pollution during totality:

- Disable your camera flash if you take pictures.
- Select a viewing site with few or no lights or signs.
- Position yourself to keep lights behind you or out of your field of vision during the eclipse.
- Move as far away from lights as possible.

States with the Best Chances

Weather, track length and duration are three factors I used in selecting states with the best chance to see the total eclipse. Oregon, Idaho, Wyoming, Nebraska, and Missouri came out on top as follows:

States with the Best Chances	Eclipse Track Miles	Average Duration / % of 2:41	Best Weather Chances*
Nebraska	468	2:34 / 95%	+/=
Wyoming	366	2:25 / 90%	+
Oregon	337	2:04 / 77%	+
Idaho	312	2:15 / 83%	+
Missouri	300	2:40 / 99%	=

*Please see Jay Anderson's excellent website Eclipsophile.com for a detailed analysis of weather expectations for this eclipse.

- Nearly 75% of the eclipse track covers these five states, almost 1,800 miles of totality.
- Most of that five state area receives more than 90% of the maximum duration of totality for this event.
- The weather forecast for August in the United States overall calls for about a 60% chance of clear skies*.

Attend an Event or Make your Own?

If there's an eclipse watching event scheduled within reasonable driving distance, consider attending. Vendors, restrooms, parking and viewing guidance during totality are a few benefits you may get at an organized event.

What if you would like to avoid crowds and have your own event? You can select an appropriate site in a cloud free area just before eclipse day, make your own preparations and do your own thing.

Either way, don't commit to any one destination until you're sure of the weather. Even if you must settle for a site farther from the eclipse centerline, a shorter totality under clear skies is better than none at all.

Eclipse Road Trip Planning

Weather is the most important factor to consider for your eclipse viewing site. Be ready to change your travel plans in the final days before the eclipse if a weather system threatens your planned destination. Have at least three options in mind along the eclipse track and do some advance planning so you can pick out a cloud-free eclipse site on short notice and get there before the eclipse starts.

Give yourself plenty of time on eclipse day. The total solar eclipse event begins with what is called "first contact" - when the moon first starts to cover the edge of the sun - about 90 minutes before totality begins. **You want to be at your spot at least 2 hours before totality begins to see the entire eclipse. So if it takes two hours to get there, you'll be leaving home four hours before totality starts.**

Know in advance how long it takes to get to your eclipse viewing site. When estimating your travel time, don't forget to include delays for Monday morning traffic, unforeseen accidents up ahead and any summer road construction along your route. Don't be afraid to generously pad your estimated travel time.

Remember that online map driving time estimates do not include stops for refueling.

Print a map and driving directions to your planned destination. The printed map will be your backup in case of weak phone signals.

To know exactly when to leave for each of your three eclipse viewing destinations follow these steps:

1. Pick out three sites where you would like to see the total eclipse of the sun. Select one site as your primary and two backup sites along the eclipse track, one west and one east of your primary destination.

2. Record the location information and totality start times for all three sites in the worksheet below.

3. Figure out how long it will take you to drive to each site and add two hours so that you'll arrive before "first contact". Record that time below.

4. Record your "Go Time" - travel time plus two hours before totality starts - in the space provided below.

Example	Site	Totality Start Time	Padded Travel Time Plus 2 Hours	Go Time
Primary	St Joseph, MO #18	1:06 p.m.	1.5 + 2 = 3.5 hrs	9:36 a.m.
Backup West	Grand Island, NE #88	12:58 p.m.	5.5 + 2 = 7.5 hrs	5:28 a.m.
Backup East	Arrow Rock, MO #90	1:10 p.m.	2.5 + 2 = 4.5 hrs	8:40 a.m.

August 21, 2017	Site	Totality Start Time	Padded Travel Time Plus 2 Hours	Go Time
Primary				
Backup West				
Backup East				

Planning and Preparation Items

Going out to see this eclipse may be like visiting an air show, with lots of walking under sunny skies. Below are a few items to think about and prepare for the day.

Eclipse glasses and filters - Eclipse glasses for everyone in your group plus solar filters for any binoculars, cameras or scopes you are bringing. A couple of spare pairs of eclipse glasses would be nice.

Bottled water - August in the U.S. is usually hot weather. Bring enough bottled water for everyone in your group.

Cooler and ice - for the bottled water and any lunches you might want to bring.

Slotted spoon - A large slotted spoon with round holes will make a great improvised pinhole projector. Just cast the shadow of the spoon onto any flat surface.

Cash - Some event vendors only take cash for parking and admission. ATMs may not be available at all venues.

Gas - Fill up and check tire pressure well before it's time to leave on Monday morning.

Napkins - Wet wipes, paper towels, toilet paper, trash bags and plastic forks & spoons are road trip staples.

Protect yourself - A hat, sunscreen, mosquito repellent, sunglasses (not for eclipse viewing) and comfortable shoes will make the day much more pleasant.

Printed map/directions to your destination area - Especially if you're traveling to an unfamiliar spot. The paper is a backup for your GPS or smartphone if cell signals are weak or unavailable. Consider getting a road atlas for the state(s) you'll be traveling through.

Vehicle maintenance - Oil checked, windows clean, spare tire and jack in working order, all tires have good tread, all fluids checked. Eclipse day is no time to run into car trouble.

Bring this book - As a handy reference for nearby eclipse viewing sites. If a site you selected is too crowded, charges too much, is closed or is otherwise off limits you may be able to spot another nearby site in this book. **If you can get within a mile or two of the GPS coordinates listed for any site in this book, your view of the eclipse and the duration of totality will be nearly identical to the listed location.**

Price of lodging - If you are planning to stay overnight somewhere during the weekend prior to the eclipse, watch out for price gouging. Consider planning to stay a little farther away from the centerline and drive a little longer on Monday morning to avoid overpriced accommodations near the eclipse centerline.

Food - Bringing some brown bag lunches would be a great idea. Food options at eclipse watching venues may be crowded and/or overpriced. After totality, long lines will be the rule rather than the exception at nearby eateries. So relax and have something to eat while you recall the eclipse and wait for the crowd to thin out.

Eat light - The night before and the morning of the eclipse, eat a little lighter than normal. Eating a large dinner or a big breakfast just before your eclipse trip could result in a long wait for the rest room at the worst possible time. Plan your rest stops so that everyone is comfortable during totality.

Fees - Many organized eclipse events will be charging for parking and/or admission. Most National Park Service sites charge an admission fee. Avoid surprises and try to find out in advance if the site you plan to visit charges a fee.

Security restrictions - If you are attending an organized eclipse watching event, find out in advance what they will allow or not allow in terms of pets, coolers, bottles, cans, drinks, chairs and outside food.

Lawn chairs - Lounge type chairs are even better to let you look up without neck strain.

Eye Safety
What is a Solar Filter?

A solar filter is a product carefully designed, marketed and sold expressly for the purpose of protecting eyes and optics from the harmful rays of the sun. A solar filter can take many forms. Sheets or rolls of film, film or metal coated glass mounted in "cells" sized for use on binoculars and telescopes, cardboard or plastic framed eclipse glasses and solar viewers are a few of the types of solar filters available.

Different materials produce different colors and kinds of images of the sun. Eclipse glasses are usually made using black polymer and give an orange color to the image. Other types of solar films and metal coated glass may give a white or blue appearance to the solar image. Welders glass shade #14 will give a green image of the sun.

Where to Get Solar Filters Online

Check Amazon.com for solar filters and eclipse glasses
http://www.amazon.com/s?field-keywords=solar+filters+eclipse+glasses

Rainbow Symphony
https://www.rainbowsymphonystore.com

Seymour Solar
https://www.seymoursolar.com

Thousand Oaks Optical
http://thousandoaksoptical.com

American Paper Optics
http://www.eclipseglasses.com

Do not wear eclipse glasses *then* look through binoculars or a telescope. The scope or binoculars must wear the filter over the large (objective) end.

Binoculars and Telescopes Need Filters
If you do not have solar filters for your binoculars or telescope then just leave that scope or binoculars at home on eclipse day.

What is NOT a Solar Filter?

- Smoked glass
- Welding visors/glass with unknown filter rating.
- Stacked welders glass rated less than #14.
- CD or DVD disks
- X-Ray film
- Floppy disks
- Space blankets
- Mylar from packages, wrappers or balloons
- Exposed film or negatives
- Sunglasses or multiple pairs of sunglasses
- Tinted safety glasses
- Polarizing or neutral density camera filters

Expert testing[1] finds that while some of these items may reduce the glare, **none of them provide enough protection from the damaging rays of the sun.**

[1]Please visit this link for more eye safety information.
http://eclipse.gsfc.nasa.gov/SEhelp/safety2.html
EYE SAFETY DURING SOLAR ECLIPSES
B. Ralph Chou, MSc, OD
Associate Professor, School of Optometry, University of Waterloo

Do not look at any part of the exposed sun without a solar filter.

How to Use a Solar Filter

Before using a solar filter check it carefully for damage or defects. Read the instructions that came with the product and verify that this filter is appropriate for your intended use. With any solar filter you select, the image of the solar disk should not be overly bright. If the image is uncomfortably bright **immediately look away** and try to determine what the problem is. Do not use a damaged or defective solar filter.

Always put the Filter Closest to the Sun

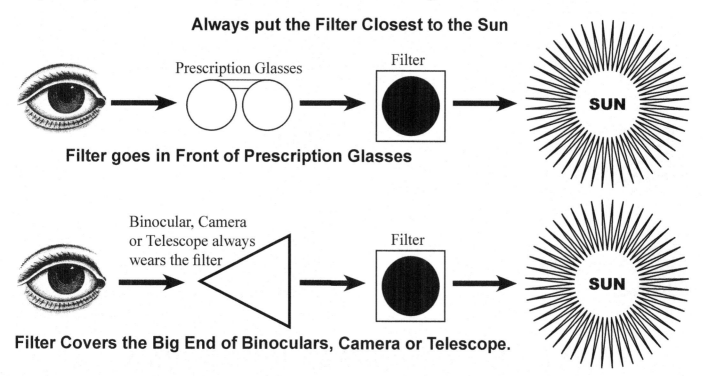

Filter goes in Front of Prescription Glasses

Filter Covers the Big End of Binoculars, Camera or Telescope.

Don't use a telescope or binoculars without a solar filter covering the objective end.

Q and A about looking at the Solar Eclipse

If the sun will be 99.9% covered can I look up unfiltered at that very thin crescent of the sun?
No. That thin crescent of exposed sun is still strong enough to cause eye damage. **Do not look at any part of the exposed sun without a solar filter.**

How do total solar eclipses work?
This NASA video with Fred Espenak explains it all.

Scan for Video

Launchpad: Solar Eclipses video from NASA

Can I look at totality without a filter?
Yes. When the sun is totally eclipsed by the moon you must look without a filter. During totality the filter would block the display of the corona around the sun.

How do I know when the sun is totally eclipsed?
When the sun has gone totally dark with no parts of the sun visible through your filter, then you can look. Before any part of the sun comes back out from behind the moon put your filter back on.

Can I look at the bright spots just before and after totality? They call it Baily's Beads and the "diamond ring". Can I look at those?
No. You cannot look with your naked eye at any part of the exposed sun. You must use eye protection during these phenomena. Watch the crescent edge of the sun get smaller and thinner through your solar filter. When the sun is totally covered by the moon and you can no longer see any part of the sun through the filter then carefully look past the filter. Only by observing the conditions where you stand will you be able to know when you can safely look at the total eclipse.

Do not look at any part of the exposed sun without a solar filter.

Eclipse Related Web Sites

 Online maps of each state for all locations shown in this book:
www.clockpress.com/eclipse

 Eclipsophile: Climatology And Weather For Celestial Events
http://eclipsophile.com/

 National Eclipse.com Eclipse Events List
http://nationaleclipse.com/events.html

 Mr. Eclipse by F. Espenak
http://www.mreclipse.com/

 Eclipse Chasers
https://www.eclipse-chasers.com/

 Great American Eclipse.com
http://www.greatamericaneclipse.com/

 NASA Eclipse site for the 2017 Eclipse
http://eclipse2017.nasa.gov

 Xavier Jubier's Interactive Eclipse Maps
http://xjubier.free.fr/

A Note on Eclipse Photography

Leave Totality Photos to the Experts

If you're like me, you've never seen a total eclipse of the sun. Totality lasts less than three minutes for this eclipse. I'm not planning to spend any of that time trying to take pictures during the very best part. I don't want to get caught fumbling with my tripod or watching the eclipse through a viewfinder or on my smartphone when I could be enjoying a natural spectacle.

There are many books and websites about imaging total solar eclipses. The links above will get you started if you would like to do that. But your first total eclipse of the sun is probably not the best time for practicing new skills. Besides, this will probably be the most photographed total solar eclipse ever with many experienced professionals producing fantastic images for everyone to see afterward.

Don't let a camera, smartphone or any technology get between you and this spectacular natural event.

Landscape Photography

In the moments just before and just after totality the landscape will be bathed in an eerie twilight that is produced by the thin crescent of the exposed solar disk. This is a great opportunity for interesting photos of all your friends and the crowd enjoying the lunar shadow and the surrounding landscape. Selecting a scenic area to view the eclipse will contribute to your enjoyment.

Practice capturing sunrise and sunset photos and know your settings well before eclipse day. The more practice you do beforehand the more you'll enjoy your day.

And remember, never point your camera at the sun without a solar filter in place. Not only can you damage your eyes looking at the sun through your camera but the camera itself can be damaged if the exposure is long.

Last but not least, **disable the flash** on your camera for eclipse day. The flash will distract everyone and diminish your ability to see during totality.

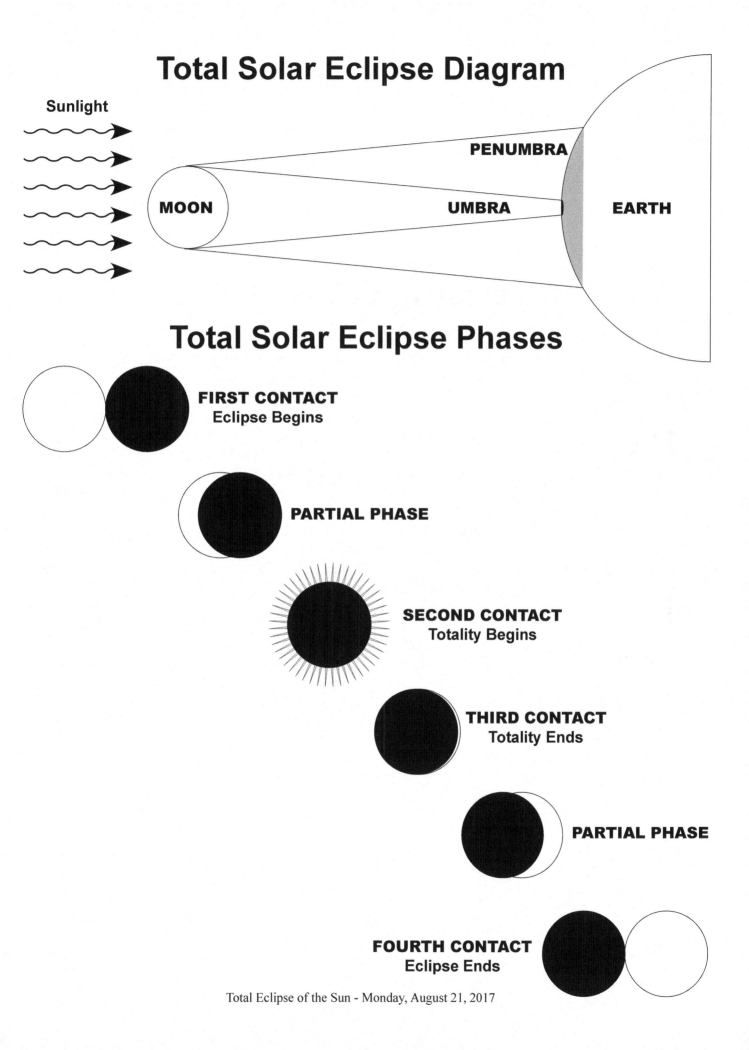

Total Solar Eclipse Diagram

Sunlight

PENUMBRA

MOON

UMBRA

EARTH

Total Solar Eclipse Phases

FIRST CONTACT
Eclipse Begins

PARTIAL PHASE

SECOND CONTACT
Totality Begins

THIRD CONTACT
Totality Ends

PARTIAL PHASE

FOURTH CONTACT
Eclipse Ends

Total Eclipse of the Sun - Monday, August 21, 2017

Glossary of Eclipse Terms

Baily's beads - Bright points of sunlight around the irregular edges of the moon seen just before totality starts and just after totality ends. First explained by Francis Baily in 1836.

center line - The path of the darkest part of the shadow cast by the moon on the earth. Totality lasts longest along the center line of the path of totality.

corona - The white glowing atmosphere of the sun not normally visible on earth but revealed during totality.

diamond ring - When only one point of light from the sun is visible along the edge of the moon just before and just after totality.

first contact - When the edge of the moon first appears to cover the edge of the sun.

fourth contact - When the edge of the moon no longer appears to cover the sun.

penumbra - Fuzzy outer shadow surrounding a darker central shadow. Eclipse watchers under the penumbral shadow will only see a partial eclipse.

second contact - When the moon fully covers the sun. Totality begins at second contact.

shadow bands - Sometimes seen just before and just after totality on light colored surfaces. May be caused by air currents in the earth's atmosphere filtering the light from the thin crescent of the solar disc.

third contact - When the sun starts to become visible at the end of totality.

totality - When the moon fully covers the sun.

umbra - The darkest central part of a shadow. Eclipse watchers under the umbral shadow will see a total eclipse of the sun.

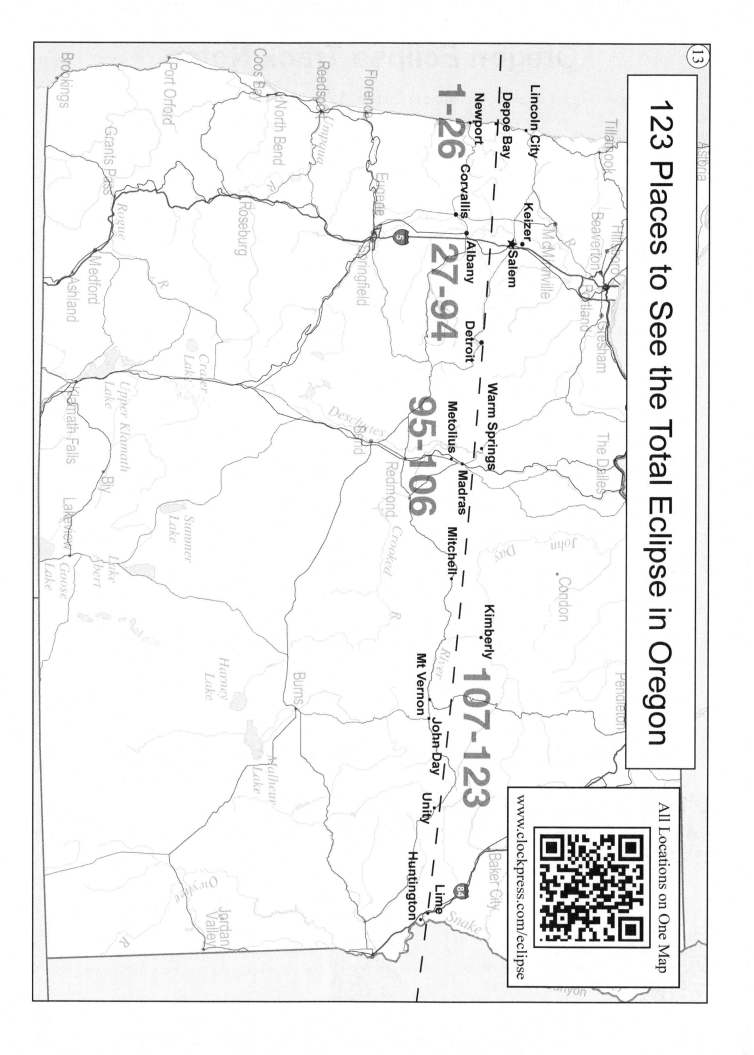

123 Places to See the Total Eclipse in Oregon

All Locations on One Map

www.clockpress.com/eclipse

Oregon Eclipse Track Notes

How many miles long is the eclipse centerline in Oregon? **337 miles**

What is the average duration of totality along the centerline in Oregon? **2:04 (m:ss)**
How long does totality last on the centerline at the Oregon west coast? **1:59 (m:ss)**
How long does totality last on the centerline at the Oregon/Idaho border? **2:10 (m:ss)**

When does the partial eclipse start on the centerline at the Oregon west coast? **9:04 a.m. (PDT)**
When does totality start on the centerline at the Oregon west coast? **10:15 a.m. (PDT)**
When does the partial eclipse end on the centerline at the Oregon west coast? **11:36 a.m. (PDT)**

When does the partial eclipse start on the centerline at the Oregon/Idaho border? **9:10 a.m. (PDT)**
When does totality start on the centerline at the Oregon/Idaho border? **10:24 a.m. (PDT)**
When does the partial eclipse end on the centerline at the Oregon/Idaho border? **11:48 a.m. (PDT)**

Weather prospects for this event in Oregon? **Best in central Oregon near Madras and eastward.**
Check local forecasts and satellite maps for the best weather information for this eclipse.

Cities and Highways in Totality by Location

1-26: Oregon Coast - **US-101** - Neskowin, Depoe Bay, Newport, Toledo

95-106: Central Oregon - **US-97, US-26** - Metolius, Madras, Terrebonne, Culver

27-80: Willamette Valley - **I-5** - Keizer, Salem, Albany, Corvallis, Lebanon

107-111: Central Oregon - **US-26** - Mitchell, Fossil, Kimberly, John Day Fossil Beds

81-94: West Central Oregon - **OR-22** - Silverton, Stayton, Lyons, Mill City, Gates, Detroit

112-123: Eastern Oregon - **US-26, I-84** - Mt. Vernon, Prairie City, Unity, Sumpter, Baker City, Huntington

Weather and Eclipse Related Links

Website

Western Oregon Weather
Portland, OR NWS Office
http://www.wrh.noaa.gov/pqr/

Northwest U.S. GOES Satellite - Visible Loop
http://www.ssd.noaa.gov/goes/east/nw/h5-loop-vis.html

Eastern Oregon Weather
Pendelton, OR NWS Office
http://www.wrh.noaa.gov/pdt/

Oregon State Parks Eclipse 2017
http://oregonstateparks.org/index.cfm?do=v.page&id=60

Featured Eclipse Destinations

15

Yaquina Head Outstanding Natural Area on the Oregon coast features Oregon's tallest lighthouse and striking eclipse viewing opportunities. Beaches, wildlife and an interpretive center are also available. Totality lasts nearly two minutes at this scenic park. Totality starts at 10:15 a.m. PDT. See Oregon #16 for location details.

Website

Photos

The Cove Palisades State Park in central Oregon offers expansive views from the cliffs surrounding Lake Billy Chinook and a myriad of outdoor activities. Eclipse watchers will find lots of unobstructed sky for the totality phase that lasts nearly two minutes at this state park. Totality starts at 10:19 a.m. PDT. See Oregon #96 for location details.

Website

Photos

State Capitol State Park in Salem has a tree called the Moon Tree*, grown from a seed carried to the surface of the Moon in 1971. This beautiful park on the state capitol grounds contains several fountains and will be an attractive place to view the total eclipse of the sun. Totality starts at 10:17 a.m. PDT. See Oregon #68 for location details.

Website

Photos

The **John Day Fossil Beds National Monument** in central Oregon has a multi-colored landscape. The banded hills color dramatically during changing light conditions and should put on an excellent show during the eclipse. Two sections of the monument, the Painted Hills Overlook and the Sheep Rock Unit, lie directly on the path of totality. Totality starts at 10:20 a.m. PDT. See Oregon #107 and #110 for location details.

Website

Photos

*Read more about the Moon Trees from NASA's website

Moon Trees
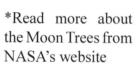

OR Crossing Points Table

| \multicolumn{6}{c}{Where the Eclipse Centerline Crosses Highways in Oregon} |

Loc#	Hwy	Nearest Mile Marker, Cross Street or Exit, City	TStart	TLasts
8	US-101	MM 125, Fishing Rock St, Depoe Bay, OR	10:15	1:59
33	99W	MM 66, Monmouth, OR	10:16	2:00
57	I-5	MM 245, Exit 244 (SB), Jefferson, OR	10:17	2:01
83	OR-22	MM 19, Kingdom Lane SE, Stayton, OR	10:17	2:01
100	US-26	MM 112, NW Fir Ln, Madras, OR	10:19	2:04
105	US-97	MM 87, NE Elm Ln, Madras, OR	10:19	2:04
109	OR-207	Cherry Ln, Mitchell, OR	10:20	2:05
112	US-395	MM 106, NF-131, Mt. Vernon, OR	10:22	2:07
114	US-26	MM 183, Prairie City, OR	10:22	2:08
116	US-26	MM 207, Munn Rd, Unity, OR	10:23	2:08
121	I-84	Exit 342, BUS US-30, Huntington, OR	10:24	2:10

Start Times = Pacific Daylight Time

Oregon Coast and Willamette Valley

Scan for Map   **Start Times = Pacific Daylight Time**

Scan for Map

9 Fogarty Creek State Park
Depoe Bay OR 97341
44.83845 -124.04884
Totality Starts at 10:15 / Lasts 1:59

1 Neskowin Creek RV Resort
50500 US-101, Neskowin, OR 97301
45.09527 -123.97876
Totality Starts at 10:16 / Lasts 1:36

10 Boiler Bay State Scenic Viewpoint
Depoe Bay, OR 97341
44.82948 -124.06485
Totality Starts at 10:15 / Lasts 1:59

2 Roads End State Park
Lincoln City, OR 97367
45.00814 -124.00853
Totality Starts at 10:16 / Lasts 1:48

11 Depoe Bay City Park
Depoe Bay, OR 97341
44.81222 -124.06261
Totality Starts at 10:15 / Lasts 2:00

3 D River State Park
Lincoln City, OR 97367
44.96748 -124.01698
Totality Starts at 10:16 / Lasts 1:52

12 Rocky Creek State Wayside
Depoe Bay, OR 97341
44.78535 -124.07306
Totality Starts at 10:15 / Lasts 1:59

4 Canyon Drive Park
801 US-101, Lincoln City, OR 97367
44.95938 -124.01941
Totality Starts at 10:16 / Lasts 1:52

13 Otter Crest State Wayside
Otter Rock, OR 97369
44.76028 -124.06624
Totality Starts at 10:15 / Lasts 1:58

5 Taft Park
Lincoln City, OR 97367
44.93079 -124.02091
Totality Starts at 10:16 / Lasts 1:55

14 Devils Punch Bowl State Park
Otter Rock, OR 97369
44.74717 -124.06445
Totality Starts at 10:15 / Lasts 1:57

6 Gleneden County Park
Gleneden Beach, OR 97388
44.88651 -124.03448
Totality Starts at 10:16 / Lasts 1:58

15 Beverly Beach State Park
Newport, OR 97365
44.72884 -124.05527
Totality Starts at 10:15 / Lasts 1:56

7 Gleneden Beach State Park
Gleneden Beach, OR 97388
44.87648 -124.03693
Totality Starts at 10:16 / Lasts 1:58

16 Yaquina Head Natural Area
750 NW Lighthouse Dr, Newport, OR
44.67631 -124.07776
Totality Starts at 10:15 / Lasts 1:50

8 CP 3420 US-101
Depoe Bay, OR 97341
44.84419 -124.04730
Totality Starts at 10:15 / Lasts 1:59

17 Agate Beach State Park
Newport, OR 97365
44.65929 -124.05671
Totality Starts at 10:15 / Lasts 1:48

Plan to be at your eclipse viewing site at least two hours before totality starts.

Scan for Map

18 **Betty Wheeler Memorial Field**
852 NW Nye St, Newport, OR 97365
44.64253 -124.05465
Totality Starts at 10:15 / Lasts 1:45

19 **Don Davis Park**
840 W Olive St, Newport, OR 97365
44.63630 -124.06346
Totality Starts at 10:15 / Lasts 1:44

20 **Mombetsu Sister City Park**
620 SW Neff Way, Newport, OR
44.63330 -124.05961
Totality Starts at 10:15 / Lasts 1:44

21 **Yaquina Bay State Park**
Newport, OR 97365
44.62366 -124.06248
Totality Starts at 10:15 / Lasts 1:42

22 **Mike Miller County Park**
South Beach, OR 97366
44.60373 -124.05132
Totality Starts at 10:15 / Lasts 1:40

23 **Newport Airport**
135 SE 84th St, Newport, OR 97366
44.58662 -124.06060
Totality Starts at 10:15 / Lasts 1:37

24 **Moonshine Park**
3870 Moonshine Park Rd, Logsden, OR
44.77756 -123.83491
Totality Starts at 10:16 / Lasts 1:59

25 **Olalla Reservoir City Park**
Toledo, OR 97391
44.68118 -123.92891
Totality Starts at 10:16 / Lasts 1:51

26 **East Slope (Kallenbach) Park**
Toledo, OR 97391
44.61252 -123.92912
Totality Starts at 10:16 / Lasts 1:42

Scan for Map

27 **Ritner Creek Park**
22300 Gage Rd, Monmouth, OR
44.74001 -123.48827
Totality Starts at 10:16 / Lasts 1:59

28 **Fort Hoskins County Park**
Philomath, OR 97370
44.67698 -123.45925
Totality Starts at 10:16 / Lasts 1:54

29 **Dallas City Park**
Academy St, Dallas, OR 97338
44.92748 -123.32216
Totality Starts at 10:16 / Lasts 1:54

30 **Nesmith County Park**
Rickreall, OR 97371
44.93100 -123.22861
Totality Starts at 10:17 / Lasts 1:53

31 **Monmouth Recreational Park**
Monmouth, OR 97361
44.85279 -123.22210
Totality Starts at 10:17 / Lasts 1:59

32 **Riverview Park**
Independence, OR 97351
44.85379 -123.18269
Totality Starts at 10:17 / Lasts 1:59

33 CP 8600 S Pacific Hwy W
Monmouth, OR 97361
44.81158 -123.22716
Totality Starts at 10:16 / Lasts 2:00

34 **Sarah Helmick State Park**
Monmouth, OR 97361
44.78200 -123.23700
Totality Starts at 10:16 / Lasts 2:01

35 **Bald Hill Natural Area**
Corvallis, OR 97330
44.56789 -123.33209
Totality Starts at 10:16 / Lasts 1:40

Plan to be at your eclipse viewing site at least two hours before totality starts.

Scan for Map

36 Adair County Park
Corvallis, OR 97330
44.67466 -123.21304
Totality Starts at 10:16 / Lasts 1:55

37 Chip Ross Park
NW Lester Ave, Corvallis, OR 97330
44.60944 -123.28467
Totality Starts at 10:16 / Lasts 1:46

38 Chepenefa Springs Park
Corvallis, OR 97330
44.59776 -123.28619
Totality Starts at 10:16 / Lasts 1:44

39 Student Legacy Park
Corvallis, OR 97331
44.56303 -123.28117
Totality Starts at 10:16 / Lasts 1:40

40 Avery Park
Corvallis, OR 97333
44.55445 -123.27278
Totality Starts at 10:16 / Lasts 1:38

41 North Riverfront Park
Corvallis, OR 97333
44.56659 -123.25341
Totality Starts at 10:16 / Lasts 1:40

42 Village Green City Park
Corvallis, OR 97330
44.59905 -123.23817
Totality Starts at 10:16 / Lasts 1:45

43 Bowers Rocks State Park
Albany, OR 97321
44.63458 -123.15674
Totality Starts at 10:17 / Lasts 1:51

44 Bryant Park
801 Bryant Way SW, Albany, OR
44.63728 -123.11425
Totality Starts at 10:17 / Lasts 1:51

45 Pioneer Park
Tangent, OR 97389
44.54461 -123.11299
Totality Starts at 10:17 / Lasts 1:38

46 Tangent City Park
Tangent, OR 97389
44.53925 -123.10927
Totality Starts at 10:17 / Lasts 1:37

47 Bass Estate Park
Tangent, OR 97389
44.53840 -123.10246
Totality Starts at 10:17 / Lasts 1:37

48 Freeway Lakes County Park
6000 Three Lakes Rd SE, Albany, OR
44.59070 -123.06273
Totality Starts at 10:17 / Lasts 1:45

49 Periwinkle Park
Albany, OR 97322
44.62046 -123.07786
Totality Starts at 10:17 / Lasts 1:49

50 Millersburg Park
Alexander Ln NE, Albany, OR 97321
44.68561 -123.06564
Totality Starts at 10:17 / Lasts 1:57

51 Linn County Fairgrounds
3700 Knox Butte Rd E, Albany, OR
44.64412 -123.05618
Totality Starts at 10:17 / Lasts 1:53

52 Timber-Linn Memorial Park
Albany, OR 97322
44.63490 -123.05500
Totality Starts at 10:17 / Lasts 1:52

53 Lexington Park
3000 21st Ave SE, Albany, OR 97322
44.62093 -123.06801
Totality Starts at 10:17 / Lasts 1:49

Plan to be at your eclipse viewing site at least two hours before totality starts.

Scan for Map

54 Century Park
Lebanon, OR 97355
44.54321 -122.91354
Totality Starts at 10:17 / Lasts 1:40

63 Clark Creek City Park
Salem, OR 97302
44.91058 -123.03896
Totality Starts at 10:17 / Lasts 1:55

Scan for Map

55 Booth Park
Lebanon, OR 97355
44.53865 -122.90064
Totality Starts at 10:17 / Lasts 1:39

64 Gilmore Field City Park
Salem, OR 97302
44.91810 -123.03314
Totality Starts at 10:17 / Lasts 1:54

56 Ankeny National Wildlife Refuge
Jefferson, OR 97352
44.79736 -123.07772
Totality Starts at 10:17 / Lasts 2:01

65 Bush's Pasture Park
600 Mission St SE, Salem, OR 97302
44.92782 -123.03653
Totality Starts at 10:17 / Lasts 1:53

57 CP I-5, OR-99E
Jefferson, OR 97352
44.80299 -123.03045
Totality Starts at 10:17 / Lasts 2:01

66 Pringle City Park
Salem, OR 97301
44.93442 -123.03741
Totality Starts at 10:17 / Lasts 1:53

58 Bryan Johnston Park
Salem, OR 97306
44.87003 -123.05280
Totality Starts at 10:17 / Lasts 1:58

67 Minto-Brown Island Park
2200 Minto Island Rd SW, Salem, OR
44.92487 -123.07277
Totality Starts at 10:17 / Lasts 1:54

59 Secor Park
Salem, OR 97306
44.87546 -123.06531
Totality Starts at 10:17 / Lasts 1:58

68 State Capitol State Park
Salem, OR 97310
44.93803 -123.02866
Totality Starts at 10:17 / Lasts 1:54

60 Sumpter School City Park
Salem, OR 97306
44.87839 -123.04529
Totality Starts at 10:17 / Lasts 1:58

69 Orchard Heights City Park
Salem, OR 97304
44.95961 -123.06208
Totality Starts at 10:17 / Lasts 1:50

61 Wendy Kroger City Park
Salem, OR 97302
44.88806 -123.05720
Totality Starts at 10:17 / Lasts 1:57

70 Grant School Park
1390 Cottage St NE, Salem, OR
44.95163 -123.02638
Totality Starts at 10:17 / Lasts 1:51

62 Hillview City Park
Salem, OR 97302
44.90005 -123.04341
Totality Starts at 10:17 / Lasts 1:57

71 Oregon State Fair Center
2330 17th St NE, Salem, OR 97303
44.95929 -123.00927
Totality Starts at 10:17 / Lasts 1:50

Plan to be at your eclipse viewing site at least two hours before totality starts.

72 Keizer Rapids Park
1900 Chemawa Rd N, Salem, OR
44.99368 -123.06091
Totality Starts at 10:17 / Lasts 1:46

73 Wilark County Park
Keizer, OR 97303
45.00979 -123.01632
Totality Starts at 10:17 / Lasts 1:44

74 Northwest University Salem
Faith Ave NE, Brooks, OR 97305
45.05473 -122.95186
Totality Starts at 10:17 / Lasts 1:37

75 Lake Labish County Park
Salem, OR 97305
45.02179 -122.96911
Totality Starts at 10:17 / Lasts 1:42

76 McKay School City Park
Salem, OR 97305
44.96159 -122.97909
Totality Starts at 10:17 / Lasts 1:50

77 Hoover School City Park
Salem, OR 97301
44.94611 -122.99459
Totality Starts at 10:17 / Lasts 1:51

78 Geer Community Park
Salem, OR 97301
44.93234 -122.99714
Totality Starts at 10:17 / Lasts 1:53

79 Cascades Gateway City Park
2100 Turner Rd SE, Salem, OR 97302
44.91215 -122.99031
Totality Starts at 10:17 / Lasts 1:55

80 Bonesteele County Park
Salem, OR 97317
44.88093 -122.93306
Totality Starts at 10:17 / Lasts 1:57

81 Oregon Garden
879 W Main St, Silverton, OR 97381
44.99500 -122.79260
Totality Starts at 10:17 / Lasts 1:45

82 Roaring River Park
42000 Fish Hatchery Dr, Scio, OR
44.62737 -122.73072
Totality Starts at 10:17 / Lasts 1:53

83 CP 18599 N Santiam Hwy SE
Stayton, OR 97383
44.78694 -122.68138
Totality Starts at 10:17 / Lasts 2:01

84 Silver Falls State Park
Silverton, OR 97381
44.86673 -122.64984
Totality Starts at 10:17 / Lasts 1:58

85 Abiqua Creek Park
Scotts Mills, OR 97375
44.95514 -122.62068
Totality Starts at 10:17 / Lasts 1:49

86 Lyons Mehama County Park
Lyons, OR 97358
44.78850 -122.61823
Totality Starts at 10:17 / Lasts 2:01

87 John Neal Memorial Park
Lyons, OR 97358
44.78287 -122.60705
Totality Starts at 10:17 / Lasts 2:01

88 North Santiam State Park
Lyons, OR 97358
44.78177 -122.58734
Totality Starts at 10:17 / Lasts 2:01

89 North Fork County Park
Lyons, OR 97358
44.80300 -122.56099
Totality Starts at 10:17 / Lasts 2:01

Plan to be at your eclipse viewing site at least two hours before totality starts.

Scan for Map

Scan for Map

90 **BLM Fisherman's Bend Rec Site**
River Rd SE, Mill City, OR 97360
44.75722 -122.51051
Totality Starts at 10:17 / Lasts 2:02

98 **Metolius City Park**
Metolius, OR 97741
44.58936 -121.17284
Totality Starts at 10:19 / Lasts 2:00

91 **Kimmel Park**
Mill City, OR 97360
44.75303 -122.46838
Totality Starts at 10:17 / Lasts 2:02

99 **PGE Pelton Park Campground**
NW Pelton Dam Rd, Madras, OR
44.68623 -121.23617
Totality Starts at 10:19 / Lasts 2:04

92 **Minto County Park**
Gates, OR 97346
44.75388 -122.39676
Totality Starts at 10:17 / Lasts 2:02

100 CP US-26
NW Warm Springs Hwy, Madras, OR
44.70587 -121.16562
Totality Starts at 10:19 / Lasts 2:04

93 **Salmon Falls County Park**
Lyons, OR 97358
44.83258 -122.36512
Totality Starts at 10:18 / Lasts 2:00

101 **Friendship Park**
Madras, OR 97741
44.63099 -121.13216
Totality Starts at 10:19 / Lasts 2:03

94 **Detroit Lake State Rec Area**
N. Santiam Hwy SE, Detroit, OR
44.72137 -122.18722
Totality Starts at 10:18 / Lasts 2:03

102 **South Park**
Madras, OR 9774
44.62451 -121.12819
Totality Starts at 10:19 / Lasts 2:03

103 **Cowden Park**
Madras, OR 97741
44.63906 -121.11906
Totality Starts at 10:19 / Lasts 2:04

Central Oregon

95 **Round Butte Overlook Park Dam**
Madras, OR 97741
44.60078 -121.27471
Totality Starts at 10:19 / Lasts 2:01

104 **Juniper Hill Park**
Madras, OR 97741
44.63809 -121.09946
Totality Starts at 10:19 / Lasts 2:04

96 **The Cove Palisades State Park**
Terrebonne, OR 97760
44.54253 -121.27288
Totality Starts at 10:19 / Lasts 1:54

105 CP US-97
The Dalles-California Hwy
Madras, OR 97741
44.69988 -121.06537
Totality Starts at 10:19 / Lasts 2:04

97 **Culver City Park**
Culver, OR 97734
44.52384 -121.20749
Totality Starts at 10:19 / Lasts 1:52

106 **Richardson's Rock Ranch**
6683 Hay Creek Rd, Madras, OR
44.73223 -120.97827
Totality Starts at 10:19 / Lasts 2:03

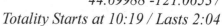

Plan to be at your eclipse viewing site at least two hours before totality starts.

Scan for Map   Scan for Map

 107 Painted Hills Overlook
John Day Fossil Beds, Kimberly, OR
44.65384 -120.25214
Totality Starts at 10:20 / Lasts 2:05

 115 Bates State Park
Bates, OR 97817
44.58836 -118.50446
Totality Starts at 10:23 / Lasts 2:05

 108 Shelton Wayside County Park
Fossil, OR 97830
44.88762 -120.07550
Totality Starts at 10:21 / Lasts 1:42

 116 CP 31011 US-26
Unity, OR 97884
44.49564 -118.24214
Totality Starts at 10:23 / Lasts 2:08

 109 CP OR-207
Mitchell, OR 97750
44.63751 -120.09812
Totality Starts at 10:20 / Lasts 2:05

 117 Sumpter Valley State Park
Sumpter, OR 97877
44.74048 -118.20439
Totality Starts at 10:23 / Lasts 1:46

 110 John Day Fossil Beds
32651 OR-19, Kimberly, OR 97848
44.55648 -119.64480
Totality Starts at 10:21 / Lasts 2:07

 118 Unity Lake State Rec Site
Unity, OR 97884
44.49134 -118.19024
Totality Starts at 10:23 / Lasts 2:08

 111 Mascall Formation Overlook
John Day Fossil Beds
44.50030 -119.62202
Totality Starts at 10:21 / Lasts 2:04

 119 Unity Forest State Park
Ironside, OR 97908
44.37027 -118.11268
Totality Starts at 10:23 / Lasts 2:06

 112 CP US-395
Mt Vernon, OR 97865
44.56498 -119.10027
Totality Starts at 10:22 / Lasts 2:07

 120 Union Creek Camp Ground
Sumpter Stage Hwy, Baker City, OR
44.68803 -118.02831
Totality Starts at 10:24 / Lasts 1:52

 113 Clyde Holliday State Rec Site
U.S 26, Mt Vernon, OR 97865
44.41638 -119.08769
Totality Starts at 10:22 / Lasts 2:01

 121 CP I-84, U.S. 30
Huntington, OR 97907
44.41304 -117.30775
Totality Starts at 10:24 / Lasts 2:10

 122 Spring Recreation Site
Huntington, OR 97907
44.37714 -117.23792
Totality Starts at 10:24 / Lasts 2:11

Eastern Oregon

 *114 CP US-26*
Prairie City, OR 97869
44.52665 -118.61563
Totality Starts at 10:22 / Lasts 2:08

123 Farewell Bend State Rec Area
23751 Old Hwy 30, Huntington, OR
44.30681 -117.22479
Totality Starts at 10:24 / Lasts 2:08

Plan to be at your eclipse viewing site at least two hours before totality starts.

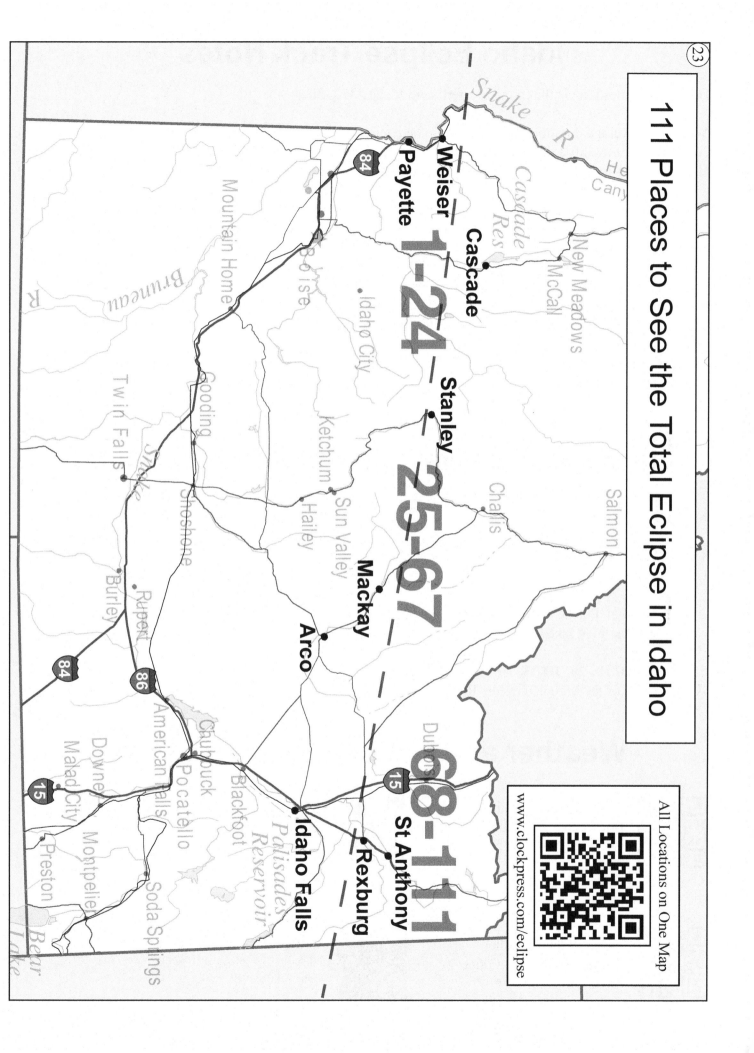

Idaho Eclipse Track Notes

How many miles long is the eclipse centerline in Idaho? **312 miles**

What is the average duration of totality along the centerline in Idaho? **2:15 (m:ss)**
How long does totality last on the centerline at the Oregon/Idaho border? **2:10 (m:ss)**
How long does totality last on the centerline at the Idaho/Wyoming border? **2:19 (m:ss)**

When does the partial eclipse start on the centerline at the Oregon/Idaho border? **10:10 a.m. (MDT)**
When does totality start on the centerline at the Oregon/Idaho border? **11:24 a.m. (MDT)**
When does the partial eclipse end on the centerline at the Oregon/Idaho border? **12:48 p.m. (MDT)**

When does the partial eclipse start on the centerline at the Idaho/Wyoming border? **10:16 a.m. (MDT)**
When does totality start on the centerline at the Idaho/Wyoming border? **11:34 a.m. (MDT)**
When does the partial eclipse end on the centerline at the Idaho/Wyoming border? **12:59 p.m. (MDT)**

Weather prospects for this event in Idaho? **Best along the Snake River Plain in the west and east.**
Check local forecasts and satellite maps for the best weather information for this eclipse.

Cities and Highways in Totality by Location

1-14: Western Idaho - **US-95** - Payette, Weiser, Midvale

15-25: Western Idaho - **ID-55** - Cascade, Ola, Banks, Garden Valley

26-57: National Forest - **ID-21, ID-75** - Lowman, Stanley, Ketchum, Mackay, Clayton

58-73: **US-93, US-20, ID-33, ID-28, I-15** - Moore, Arco, Craters of the Moon, Terreton Roberts

74-102: **I-15, US-91, US-20, US-26** - Shelley, Idaho Falls, Rigby, Ririe, Menan Rexburg, Sugar City, St. Anthony

103-111: Eastern Idaho, **ID-33, US-26** - Tetonia, Driggs, Victor, Irwin

Weather and Eclipse Related Links

Website

Western Idaho Weather
Boise, ID NWS Office
http://www.wrh.noaa.gov/boi/

Northwest U.S. GOES Satellite - Visible Loop
http://www.ssd.noaa.gov/goes/east/nw/h5-loop-vis.html

Eastern Idaho Weather
Pocatello, ID NWS Office
http://www.wrh.noaa.gov/pih/

College of Southern Idaho Eclipse Site
https://herrett.csi.edu/astronomy/observatory/2017_
eclipse_FAQ.asp

Featured Eclipse Destinations

Mann Creek Reservoir off I-95 in western Idaho is near the centerline of the eclipse and will see well over two minutes of totality. Restrooms, camping, boating and fishing are available. Totality starts at 11:25 a.m. MDT. See Idaho #3 for location details.

Website

Photos

Hell's Half Acre Lava Trail System is west of Idaho Falls on US 20. This sixty thousand acre lava flow has been developed by the National Park Service to include three trails and would make an interesting eclipse viewing site. Totality starts at 11:32 a.m. MDT. See Idaho #70 for location details.

Website

Photos

Lake Cascade State Park north of Boise offers multiple campgrounds, boat ramps and day use areas for outdoor activities. Nearly two minutes of totality makes this a great place to enjoy eclipse day. Totality starts at 11:26 a.m. MDT. See Idaho #17 for location details.

Website

Photos

North Menan Butte National Natural Landmark is an 800 foot tall volcanic tuff cone just north of Idaho Falls. Experience two minutes and eighteen seconds of totality overlooking the Snake River plain from the top of this rare formation. Totality starts at 11:32 a.m. MDT. See Idaho #89 for location details.

Website

Photos

ID Crossing Points Table

Where the Eclipse Centerline Crosses Highways in Idaho

Start Times = Mountain Daylight Time

Loc#	Hwy	Nearest Mile Marker, Cross Street or Exit, City	TStart	TLasts
4	US-95	MM 94, Mann Creek Rd, Weiser, ID	11:25	2:10
21	ID-55	Between MM 96-97, south of Smiths Ferry Dr, Cascade, ID	11:26	2:12
28	ID-21	Between MM 97-98, Lowman, ID	11:27	2:13
39	ID-75	Where I-75 crosses the Salmon River, Stanley, ID	11:28	2:14
58	US-93	About 3 miles south of Trail Creek Rd, Mackay, ID	11:30	2:15
68	ID-22	6.6 miles south of ID-28, Howe, ID	11:31	2:17
69	ID-28	N 800 E, about 13 miles south of ID-22, Monteview, ID	11:32	2:17
72	ID-28/33	N 1700 E, Terreton, ID	11:32	2:18
73	I-15	About .5 mile south of exit 143, Roberts, ID	11:32	2:18
91	US-20	MM 329, south of Rexburg, ID	11:33	2:18
106	ID-33	W 4000 S, Victor, ID	11:34	2:19

Western Idaho

Scan for Map   **Start Times = Mountain Daylight Time**

1 CP Olds Ferry Rd
Snake River, Weiser, ID 83672
44.40518 -117.22336
Totality Starts at 11:24 / Lasts 2:10

2 Rest Stop on US-95
Midvale, ID 83645
44.44529 -116.79998
Totality Starts at 11:25 / Lasts 2:07

3 Mann Creek Reservoir
1699 Monroe Creek Rd
Weiser, ID 83672
44.38846 -116.89695
Totality Starts at 11:25 / Lasts 2:10

4 CP US-95
Weiser, ID 83672
44.37197 -116.87332
Totality Starts at 11:25 / Lasts 2:10

5 Monroe Creek
Campground & RV Park
822 US-95, Weiser, ID 83672
44.27052 -116.94485
Totality Starts at 11:25 / Lasts 2:08

6 Indianhead Motel & RV Park
747 E Indianhead Rd, Weiser, ID 83672
44.25982 -116.96004
Totality Starts at 11:25 / Lasts 2:07

7 Memorial Park
Gray Ave, Weiser, ID 83672
44.25418 -116.96436
Totality Starts at 11:25 / Lasts 2:06

8 Time's Square Park
State St, Weiser, ID 83672
44.24848 -116.96839
Totality Starts at 11:25 / Lasts 2:06

Scan for Map

9 Weiser Sports Complex
1180 E Park St, Weiser, ID 83672
44.25023 -116.95055
Totality Starts at 11:25 / Lasts 2:06

10 Centennial Park
Payette, ID 83661
44.09397 -116.93984
Totality Starts at 11:25 / Lasts 1:44

11 Payette Municipal Airport (North)
Payette, ID 83661
44.09804 -116.90616
Totality Starts at 11:25 / Lasts 1:46

12 CP Weiser River Trail
Weiser, ID 83672
44.36430 -116.79379
Totality Starts at 11:25 / Lasts 2:11

13 Crane Creek Reservoir
S Crane Rd, Midvale, ID 83645
44.36435 -116.55468
Totality Starts at 11:25 / Lasts 2:10

14 Ben Ross Reservoir
Swisher Rd, Indian Valley, ID 83632
44.51453 -116.44824
Totality Starts at 11:26 / Lasts 1:58

15 Antelope Campground
Forest Rd 614C, Ola, ID 83657
44.33520 -116.18667
Totality Starts at 11:26 / Lasts 2:11

16 Sage Hen Creek Campground
Forest Rd 614E, Ola, ID 83657
44.33374 -116.17866
Totality Starts at 11:26 / Lasts 2:11

17 Lake Cascade State Park
970 Dam Rd, Cascade, ID 83611
44.52094 -116.04976
Totality Starts at 11:26 / Lasts 1:52

Plan to be at your eclipse viewing site at least two hours before totality starts.

Scan for Map

18 Arrowhead RV Park
955 ID-55, Cascade, ID 83611
44.51168 -116.03772
Totality Starts at 11:26 / Lasts 1:54

Central Idaho

19 Horsethief Reservoir State Park
Cascade, ID 83611
44.51195 -115.91925
Totality Starts at 11:27 / Lasts 1:52

27 Bull Trout Lake Campground
Lowman, ID 83637
44.30256 -115.25908
Totality Starts at 11:27 / Lasts 2:08

20 Clear Creek Station
RV Park & Campground
10694 ID-55, Cascade, ID 83611
44.40760 -115.99942
Totality Starts at 11:26 / Lasts 2:05

28 CP ID-21
Lowman, ID 83637
44.20330 -115.24809
Totality Starts at 11:27 / Lasts 2:13

21 CP ID-55
Cascade, ID 83611
44.29291 -116.08271
Totality Starts at 11:26 / Lasts 2:12

29 Camp Bradley
NFD Road 158, Stanley, ID 83278
44.39705 -115.14953
Totality Starts at 11:28 / Lasts 1:58

22 Big Eddy Campground
ID-55, Cascade, ID 83611
44.22061 -116.10698
Totality Starts at 11:26 / Lasts 2:12

30 Grandjean Campground
Forest Rd 524, Lowman, ID 83637
44.14813 -115.15264
Totality Starts at 11:27 / Lasts 2:14

23 Canyon Campground
ID-55, ID 83611
44.18788 -116.11522
Totality Starts at 11:26 / Lasts 2:09

31 Sheep Trail Campground
Stanley, ID 83278
44.30615 -115.05623
Totality Starts at 11:28 / Lasts 2:06

24 Swinging Bridge Campground
ID-55, Banks, ID 83602
44.17277 -116.12059
Totality Starts at 11:26 / Lasts 2:08

32 Stanley Lake Inlet Campground
Stanley, ID 83278
44.24471 -115.06458
Totality Starts at 11:28 / Lasts 2:11

25 Cozy Cove Campground
Deadwood Reservoir
Forest Rd 555F, Garden Valley, ID 83622
44.29135 -115.65294
Totality Starts at 11:27 / Lasts 2:11

33 Stanley Lake Campground
Stanley, ID 83278
44.24884 -115.05443
Totality Starts at 11:28 / Lasts 2:11

26 Helende Campground
Forest Rd 025JB, Lowman, ID 83637
44.09210 -115.47575
Totality Starts at 11:27 / Lasts 2:08

34 Iron Creek Campground
Lowman, ID 83637
44.19879 -115.00969
Totality Starts at 11:28 / Lasts 2:13

Plan to be at your eclipse viewing site at least two hours before totality starts.

Scan for Map

35 Mountain Village Resort
3 Eva Falls Ave, Stanley, ID 83278
44.21771 -114.93118
Totality Starts at 11:28 / Lasts 2:12

44 Redfish Lake Inlet Campground
Lowman, ID 83637
44.10005 -114.95342
Totality Starts at 11:28 / Lasts 2:14

36 Sunny Gulch Campground
ID-75 Stanley, ID 83278
44.17639 -114.90950
Totality Starts at 11:28 / Lasts 2:13

45 Pettit Lake Campground
Forest Rd 362, Ketchum, ID 83340
43.98520 -114.86839
Totality Starts at 11:28 / Lasts 2:04

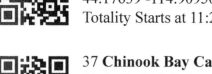

37 Chinook Bay Campground
Little Redfish Lake, Stanley, ID 83278
44.16401 -114.90422
Totality Starts at 11:28 / Lasts 2:14

46 Salmon River Campground
Stanley, ID 83278
44.25056 -114.86932
Totality Starts at 11:28 / Lasts 2:10

38 Mountain View Campground
Little Redfish Lake, Stanley, ID 83278
44.16185 -114.90500
Totality Starts at 11:28 / Lasts 2:14

47 Casino Creek Campground
Stanley, ID 83278
44.25408 -114.85643
Totality Starts at 11:28 / Lasts 2:09

39 CP ID-75
Stanley, ID 83278
44.16226 -114.88465
Totality Starts at 11:28 / Lasts 2:14

48 Riverside Campground
Stanley, ID 83278
44.26686 -114.84983
Totality Starts at 11:28 / Lasts 2:08

40 Glacier View Campground
Redfish Lake Rd, Stanley, ID 83278
44.14688 -114.91491
Totality Starts at 11:28 / Lasts 2:14

49 Mormon Bend Campground
Stanley, ID 83278
44.26170 -114.84190
Totality Starts at 11:28 / Lasts 2:08

41 Redfish Outlet Campground
Redfish Lake, Stanley, ID 83278
44.14387 -114.91252
Totality Starts at 11:28 / Lasts 2:14

50 Basin Creek Campground
Stanley, ID 83278
44.26339 -114.81961
Totality Starts at 11:28 / Lasts 2:08

42 Point Campground
Lowman, ID 83637
44.13946 -114.92495
Totality Starts at 11:28 / Lasts 2:14

51 Dutchman Flat Campground
Stanley, ID 83278
44.26572 -114.72094
Totality Starts at 11:28 / Lasts 2:07

43 Mount Heyburn Campground
Lowman, ID 83637
44.13503 -114.91540
Totality Starts at 11:28 / Lasts 2:14

52 Upper O'Brien Campground
Robinson Bar Rd, Stanley, ID 83278
44.25930 -114.69834
Totality Starts at 11:28 / Lasts 2:07

Plan to be at your eclipse viewing site at least two hours before totality starts.

Scan for Map Scan for Map

53 Marshall Creek Campground
Stanley, ID 83278
44.25418 -114.68564
Totality Starts at 11:28 / Lasts 2:08

62 Mackay Airport
Mackay, ID 83251
43.91188 -113.60507
Totality Starts at 11:30 / Lasts 2:14

54 Rest Stop ID-75
Clayton, ID 83227
44.25574 -114.59410
Totality Starts at 11:28 / Lasts 2:07

63 Bear Creek Campground
Mackay, ID 83251
43.98746 -113.45867
Totality Starts at 11:30 / Lasts 2:16

55 Holman Creek Campground
Clayton, ID 83227
44.24916 -114.53000
Totality Starts at 11:28 / Lasts 2:07

64 Moose Crossing RV Park
3798 US-93, Moore, ID 83255
43.86232 -113.45246
Totality Starts at 11:30 / Lasts 2:12

56 Sunbeam Village Resort LLC
Clayton, ID 83227
44.25912 -114.39959
Totality Starts at 11:29 / Lasts 2:05

65 Mountain View RV Park
705 W Grand Ave, Arco, ID 83213
43.62718 -113.30551
Totality Starts at 11:31 / Lasts 1:39

57 Park Creek Campground
Forest Rd 156, Mackay, ID 83251
43.83670 -114.25793
Totality Starts at 11:29 / Lasts 1:54

66 Bottolfsen Park
Arco, ID 83213
43.63225 -113.30398
Totality Starts at 11:31 / Lasts 1:40

58 CP US-93
Mackay, ID 83251
44.03178 -113.79390
Totality Starts at 11:30 / Lasts 2:15

67 Craters of the Moon / Arco KOA
2424 3000 W, Arco, ID 83213
43.62713 -113.29581
Totality Starts at 11:31 / Lasts 1:39

59 Joe T. Fallini Campground
Mackay Reservoir, Mackay, ID 83251
43.96088 -113.68335
Totality Starts at 11:30 / Lasts 2:17

Eastern Idaho

68 CP ID-22
Howe, ID 83244
43.90161 -112.78971
Totality Starts at 11:31 / Lasts 2:17

60 Mackay Tourist Park
Vadan St, Mackay, ID 83251
43.91259 -113.62222
Totality Starts at 11:30 / Lasts 2:14

69 CP ID-28
1698 N Salmon Hwy, Monteview, ID
43.86764 -112.53897
Totality Starts at 11:32 / Lasts 2:17

61 River Park Golf and RV Park
717 Capitol Ave, Mackay, ID 83251
43.91099 -113.61953
Totality Starts at 11:30 / Lasts 2:14

Plan to be at your eclipse viewing site at least two hours before totality starts.

Scan for Map

 70 Hell's Half Acre Lava Walk
W Arco Hwy, Idaho Falls, ID 83402
43.55382 -112.44199
Totality Starts at 11:32 / Lasts 1:49

 71 Mud Lake
Terreton, ID 83450
43.87196 -112.40900
Totality Starts at 11:32 / Lasts 2:17

 72 CP ID-28
1749 E 1500 N, Terreton, ID 83450
43.84133 -112.34808
Totality Starts at 11:32 / Lasts 2:18

 73 CP I-15
Veterans Memorial Hwy
Roberts, ID 83444
43.81907 -112.18854
Totality Starts at 11:32 / Lasts 2:18

 74 Shelley Park
N State St, Shelley, ID 83274
43.38425 -112.12203
Totality Starts at 11:33 / Lasts 1:10

 75 Reinhart Park
1055 Washburn Ave
Idaho Falls, ID 83402
43.50590 -112.07072
Totality Starts at 11:32 / Lasts 1:50

 76 Civitan Park
Idaho Falls, ID 83402
43.50345 -112.04394
Totality Starts at 11:32 / Lasts 1:50

 77 Central Park
400 N Holmes Ave
Idaho Falls, ID 83401
43.50069 -112.02265
Totality Starts at 11:33 / Lasts 1:50

 78 Lincoln Park
2280 Lincoln Rd, Idaho Falls, ID 83401
43.51032 -111.98852
Totality Starts at 11:33 / Lasts 1:53

 79 Russ Freeman Park
Idaho Falls, ID 83402
43.51499 -112.05090
Totality Starts at 11:32 / Lasts 1:52

 80 Idaho Falls Dog Park
Idaho Falls, ID 83402
43.52432 -112.05836
Totality Starts at 11:32 / Lasts 1:53

 81 Killdeer Park
W Commons Rd, Idaho Falls, ID 83401
43.56119 -112.03269
Totality Starts at 11:32 / Lasts 2:00

 82 Iona Park
Rainbow Ln, Iona, ID 83427
43.52889 -111.92798
Totality Starts at 11:33 / Lasts 1:58

 83 Blacktail Day Use Area
Rigby, ID 83442
43.50406 -111.76270
Totality Starts at 11:33 / Lasts 1:58

 84 Juniper Campground
Meadow Creek Rd Ririe, ID 83443
43.57985 -111.73541
Totality Starts at 11:33 / Lasts 2:10

 85 Ririe Dam Picnic Area
Ririe, ID 83443
43.58333 -111.74828
Totality Starts at 11:33 / Lasts 2:10

 86 Teton RV Park
121 E 825 N, Rigby, ID 83442
43.64576 -111.93270
Totality Starts at 11:33 / Lasts 2:14

87 Squealers Fun Park
439 N 4000 E, Rigby, ID 83442
43.68822 -111.90695
Totality Starts at 11:33 / Lasts 2:18

Plan to be at your eclipse viewing site at least two hours before totality starts.

Scan for Map

88 Jefferson County Lake
490 N 4000 E, Rigby, ID 83442
43.69886 -111.90458
Totality Starts at 11:33 / Lasts 2:18

89 North Menan Butte
National Natural Landmark
1098 N 3600 E, Menan, ID 83434
43.785589 -111.990332
Totality Starts at 11:32 / Lasts 2:18

90 Yellowstone Bear World
6010 S 4300 W, Rexburg, ID 83440
43.73905 -111.86523
Totality Starts at 11:33 / Lasts 2:19

91 CP US-20
Rexburg, ID 83440
43.76925 -111.83770
Totality Starts at 11:33 / Lasts 2:18

92 Wakeside Lake RV Park
2245 S 2000 W, Rexburg, ID 83440
43.79423 -111.82152
Totality Starts at 11:33 / Lasts 2:18

93 Thomas E. Ricks Gardens
BYU Rexburg, ID 83440
43.81561 -111.78172
Totality Starts at 11:33 / Lasts 2:17

94 Porter Park
Rexburg, ID 83440
43.82296 -111.79155
Totality Starts at 11:33 / Lasts 2:16

95 Hidden Valley Park
Palmer Cir, Rexburg, ID 83440
43.82473 -111.76502
Totality Starts at 11:33 / Lasts 2:16

96 Smith Park
Rexburg, ID 83440
43.82777 -111.77349
Totality Starts at 11:33 / Lasts 2:16

Scan for Map

97 Riverside Park
N Center St, Rexburg, ID 83440
43.83222 -111.78499
Totality Starts at 11:33 / Lasts 2:16

98 Dragons Field - Rexburg Park
N 5th W, Rexburg, ID 83440
43.83549 -111.79962
Totality Starts at 11:33 / Lasts 2:16

99 Eagle Park Campground
Rexburg, ID 83440
43.83934 -111.79822
Totality Starts at 11:33 / Lasts 2:16

100 Smith Park
Railroad Ave, Sugar City, ID 83448
43.87033 -111.75646
Totality Starts at 11:33 / Lasts 2:13

101 Heritage Park
Sugar City, ID 83448
43.87243 -111.74443
Totality Starts at 11:33 / Lasts 2:12

102 Egin Lake Campground
St Anthony, ID 83445
43.96251 -111.85336
Totality Starts at 11:33 / Lasts 2:05

103 Spring Creek Scenic Overlook
ID-33 Tetonia, ID 83452
43.82516 -111.23909
Totality Starts at 11:34 / Lasts 2:10

104 BYUI Outdoor Learning Center
Tetonia, ID 83452
43.86521 -111.13381
Totality Starts at 11:34 / Lasts 2:05

105 Driggs-Reed Memorial Airport
253 Warbird Lane, Driggs, ID 83422
43.73616 -111.11080
Totality Starts at 11:34 / Lasts 2:17

Plan to be at your eclipse viewing site at least two hours before totality starts.

Scan for Map

106 CP ID-33
Victor, ID 83455
43.66216 -111.11056
Totality Starts at 11:34 / Lasts 2:19

107 **Teton Valley RV Park**
1208 ID-31, Victor, ID 83455
43.59883 -111.12714
Totality Starts at 11:34 / Lasts 2:21

108 **Victor City Park**
Victor, ID 83455
43.60452 -111.11084
Totality Starts at 11:34 / Lasts 2:21

109 **Pioneer Park**
100-146 Elm St, Victor, ID 83455
43.59920 -111.10565
Totality Starts at 11:34 / Lasts 2:21

110 **Teton Springs Lodge & Spa**
10 Warm Creek Ln, Victor, ID 83455
43.57341 -111.10835
Totality Starts at 11:34 / Lasts 2:19

111 **Riverside Park Campground**
Forest Rd 963, Irwin, ID 83428
43.34236 -111.20550
Totality Starts at 11:34 / Lasts 1:44

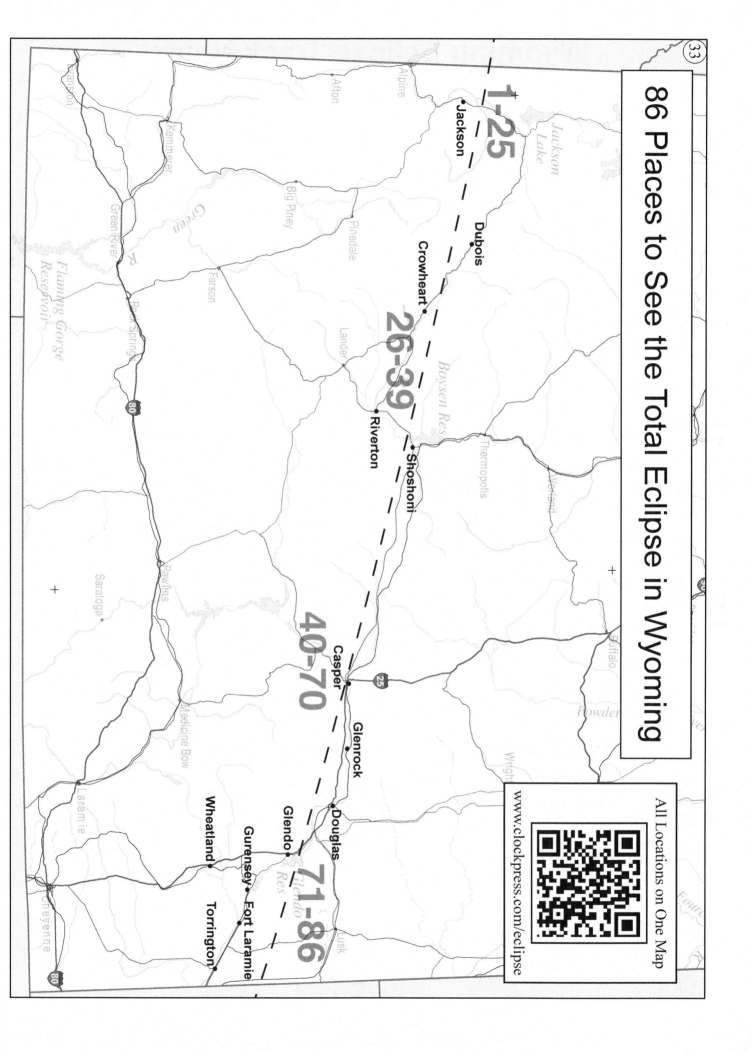

86 Places to See the Total Eclipse in Wyoming

All Locations on One Map

www.clockpress.com/eclipse

1-25

26-39

40-70

71-86

Jackson

Dubois

Crowheart

Riverton

Shoshoni

Casper

Glenrock

Douglas

Glendo

Gurensey

Fort Laramie

Wheatland

Torrington

Wyoming Eclipse Track Notes

How many miles long is the eclipse centerline in Wyoming? **366 miles**

What is the average duration of totality along the centerline in Wyoming? **2:25 (m:ss)**
How long does totality last on the centerline at the Idaho/Wyoming border? **2:20 (m:ss)**
How long does totality last on the centerline at the Wyoming/Nebraska border? **2:29 (m:ss)**

When does the partial eclipse start on the centerline at the Idaho/Wyoming border? **10:16 a.m. (MDT)**
When does totality start on the centerline at the Idaho/Wyoming border? **11:34 a.m. (MDT)**
When does the partial eclipse end on the centerline at the Idaho/Wyoming border? **12:59 p.m. (MDT)**

When does the partial eclipse start on the centerline at the Wyoming/Nebraska border? **10:25 a.m. (MDT)**
When does totality start on the centerline at the Wyoming/Nebraska border? **11:46 a.m. (MDT)**
When does the partial eclipse end on the centerline at the Wyoming/Nebraska border? **1:14 p.m. (MDT)**

Weather prospects for this event in Wyoming? **Best in central and eastern Wyoming.**
Check local forecasts and satellite maps for the best weather information for this eclipse.

Cities and Highways in Totality by Location

1-25: Western Wyoming - **US-191 -** Jackson, Moose, Kelly, Grand Teton National Park

42-70: Eastern Wyoming - **US-26, I-25 -** Casper, Evansville, Glenrock

26-41: Central Wyoming - **US-26 -** Dubois, Riverton, Shoshoni, Hells Half Acre

71-86: Eastern Wyoming - **I-25, US-26 -** Douglas, Orin, Glendo, Guernsey, Fort Laramie, Torrington

Weather and Eclipse Related Links

Website

Western Wyoming Weather
Riverton, WY NWS Office
http://www.weather.gov/riw/

Northern Plains
U.S. GOES Satellite - Visible Loop
http://www.ssd.noaa.gov/goes/east/np/h5-loop-vis.html

Eastern Wyoming Weather
Cheyenne, WY NWS Office
http://www.weather.gov/cys/

Casper Wyoming Eclipse Site
http://www.eclipsecasper.com

Featured Eclipse Destinations

Grand Teton National Park is 13 miles north of Jackson. Wildlife, boating, fishing, rafting, hiking and spectacular scenery are a few of the attractions at this national park. South of Jenny lake totality lasts longest, up to two minutes and twenty seconds outside of Jackson, Wyoming. There are numerous scenic turnouts and overlooks surrounding the park. Totality starts at 11:35 a.m. MDT. See Wyoming #4 for location details.

Website

Photos

Edness Kimball Wilkins State Park is on US-26 east of Casper. Multiple picnic areas, fishing, boating, playgrounds and restrooms are a few of the features at this state operated day-use park. Two minutes and twenty four seconds of totality will greet eclipse watchers here on the North Platte River. Totality starts at 11:42 a.m. MDT. See Wyoming #68 for location details.

Website

Photos

Ayres Natural Bridge Park is about 5 miles south of I-25 off exit 151 in eastern Wyoming. Fishing, hiking and spectacular scenery are featured at this popular attraction. Nearly two and a half minutes of totality viewed from the natural bridge make this an exciting eclipse viewing destination. Totality starts at 11:43 a.m. MDT. See Wyoming #71 for location details.

Website

Photos

Glendo State Park features boating, fishing, camping and is one of Wyoming's most popular parks. The centerline of the path of totality cuts right across this middle of this large park giving viewers two minutes twenty nine seconds of totality. Totality starts at 11:45 a.m. MDT. See Wyoming #75 for location details.

Website

Photos

WY Crossing Points Table

Where the Eclipse Centerline Crosses Highways in Wyoming

Start Times = Mountain Daylight Time

Loc#	Hwy	Nearest Mile Marker, Cross Street or Exit, City	TStart	TLasts
16	US-191	E Airport Rd, Jackson, WY	11:34	2:20
29	US-26/287	Burris-Lenore Rd, Crowheart, WY	11:37	2:22
36	US-26	Riggs Rd, WY-134, Shoshoni, WY	11:39	2:24
50	WY-220	Fremont Ave, Casper, WY	11:42	2:26
52	WY-258	MM 15, Casper, WY	11:42	2:27
69	WY-253	About 6 miles south of I-25 on WY-253	11:42	2:27
74	I-25	Between MM 115-116, Glendo Reservoir, Glendo, Wy	11:45	2:28
78	WY-270	MM 513, Glendo, WY	11:45	2:29
81	US-85	MM 123, Torrington, WY	11:46	2:29

Western Wyoming

Scan for Map   **Start Times = Mountain Daylight Time**

9 Windy Point Turnout
Moose, WY 83012
43.67729 -110.72473
Totality Starts at 11:35 / Lasts 2:17

1 Signal Mountain Campground
Alta, WY 83414
43.84072 -110.61692
Totality Starts at 11:35 / Lasts 1:57

10 Blacktail Ponds Overlook
Moose, WY 83012
43.66676 -110.69660
Totality Starts at 11:35 / Lasts 2:18

2 Deadman Point, Jackson Lake
Spalding Bay Rd, Alta, WY 83414
43.83710 -110.68142
Totality Starts at 11:35 / Lasts 2:00

11 Death Canyon Trailhead
Alta, WY 83414
43.65622 -110.78154
Totality Starts at 11:34 / Lasts 2:19

3 Potholes Turnout
Moose, WY 83012
43.80678 -110.62813
Totality Starts at 11:35 / Lasts 2:03

12 Atherton Creek Campground
Bridger Teton National Forest
Forest Rd 30374, Kelly, WY 83011
43.63754 -110.52415
Totality Starts at 11:35 / Lasts 2:18

4 Grand Teton National Park
Teton Park Rd, Moose, WY 83012
43.80171 -110.68304
Totality Starts at 11:35 / Lasts 2:05

13 Laurance S. Rockefeller Preserve
Jackson, WY 83001
43.62630 -110.77614
Totality Starts at 11:34 / Lasts 2:20

5 Jenny Lake Trailhead
Grand Teton National Park
Alta, WY 83414
43.75210 -110.72254
Totality Starts at 11:35 / Lasts 2:11

14 Gros Ventre Campground
Jackson, WY 83001
43.61623 -110.66579
Totality Starts at 11:35 / Lasts 2:20

6 Lupine Meadows Trailhead
Lupine Meadows Rd, Alta, WY 83414
43.73472 -110.74147
Totality Starts at 11:34 / Lasts 2:13

15 Jackson Hole Airport
1250 E Airport Rd, Jackson, WY
43.60535 -110.73860
Totality Starts at 11:34 / Lasts 2:20

7 Teton Glacier Turnout
Moose, WY 83012
43.71320 -110.72805
Totality Starts at 11:35 / Lasts 2:14

16 CP US-89
Jackson, WY 83001
43.60306 -110.72378
Totality Starts at 11:34 / Lasts 2:20

8 Taggart Lake Trailhead
Alta, WY 83414
43.69325 -110.73309
Totality Starts at 11:35 / Lasts 2:16

17 Teton Village
McCollister Dr, Teton Village, WY
43.58729 -110.82599
Totality Starts at 11:34 / Lasts 2:21

Plan to be at your eclipse viewing site at least two hours before totality starts.

18 Grand Teton Park Entrance
Jackson, WY 83001
43.54669 -110.73257
Totality Starts at 11:34 / Lasts 2:21

Central Wyoming

Scan for Map

26 Green River Lake Campground
Forest Rd 650, Cora, WY 82925
43.31405 -109.85836
Totality Starts at 11:36 / Lasts 2:16

19 National Museum of Wildlife Art
2820 Rungius Rd, Jackson, WY 83001
43.52086 -110.74766
Totality Starts at 11:34 / Lasts 2:20

27 Dubois / Wind River KOA
225 W Welty St, Dubois, WY 82513
43.53283 -109.63571
Totality Starts at 11:36 / Lasts 2:17

20 East Gros Ventre Butte
Saddle Butte Dr, Jackson, WY 83001
43.48458 -110.78047
Totality Starts at 11:34 / Lasts 2:16

28 CP Dinwoody Lake
Crowheart, WY 82512
43.38284 -109.36190
Totality Starts at 11:37 / Lasts 2:22

21 Miller Park
255 W Deloney Ave, Jackson, WY
43.48139 -110.76569
Totality Starts at 11:34 / Lasts 2:16

29 CP US-287
Crowheart, WY 82512
43.36969 -109.28425
Totality Starts at 11:37 / Lasts 2:22

22 Jackson Town Square
Jackson, WY 83001
43.47990 -110.76191
Totality Starts at 11:34 / Lasts 2:16

30 Ocean Lake
299 Ocean Lake Rd, Riverton, WY
43.16208 -108.61898
Totality Starts at 11:38 / Lasts 2:23

23 Virginian RV Park
750 W Broadway, Jackson, WY 83001
43.47394 -110.77905
Totality Starts at 11:34 / Lasts 2:16

31 Riverton Regional Airport
4800 Airport Rd, Riverton, WY 82501
43.06619 -108.46203
Totality Starts at 11:38 / Lasts 2:17

24 Powderhorn Park
Jackson, WY 83001
43.47192 -110.78553
Totality Starts at 11:34 / Lasts 2:15

32 Jaycee Park
W Sunset Dr, Riverton, WY 82501
43.03489 -108.41235
Totality Starts at 11:38 / Lasts 2:14

25 Jackson Hole / Snake River KOA
9705 US-89, Jackson, WY 83001
43.34138 -110.72407
Totality Starts at 11:35 / Lasts 1:58
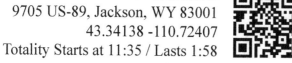

33 Rein Park Soccer Complex
W Monroe Ave, Riverton, WY 82501
43.01829 -108.40666
Totality Starts at 11:39 / Lasts 2:12

Plan to be at your eclipse viewing site at least two hours before totality starts.

Scan for Map

 34 Sunset Park
Riverton, WY 82501
43.03307 -108.39470
Totality Starts at 11:39 / Lasts 2:14

 35 Riverton City Park
Riverton, WY 82501
43.02476 -108.37962
Totality Starts at 11:39 / Lasts 2:14

 36 CP WY-789
Shoshoni, WY 82649
43.18771 -108.24370
Totality Starts at 11:39 / Lasts 2:24

 37 Boysen State Park
Shoshoni, WY 82649
43.33045 -108.15347
Totality Starts at 11:39 / Lasts 2:14

 38 Shoshoni Municipal Airport
Shoshoni, WY 82649
43.24979 -108.11947
Totality Starts at 11:39 / Lasts 2:20

 39 Wyoming Heritage Trail
Shoshoni, WY 82649
43.23244 -108.11935
Totality Starts at 11:39 / Lasts 2:21

 40 Highway Rest Area
44392 US-20, Casper, WY 82604
43.07489 -107.24499
Totality Starts at 11:40 / Lasts 2:22

41 Hells Half Acre
U.S. 26. / Hells Half Acre Rd,
Casper, WY 82604
43.04714 -107.09257
Totality Starts at 11:41 / Lasts 2:22

42 Casper–Natrona Airport
8500 Airport Pkwy, Casper, WY
42.90338 -106.46280
Totality Starts at 11:42 / Lasts 2:24

Scan for Map

 43 Buckboard Park
Buckboard Rd, Casper, WY 82604
42.81517 -106.42002
Totality Starts at 11:42 / Lasts 2:27

 44 Casper KOA
1101 Prairie Ln, Bar Nunn, WY 82601
42.91427 -106.33971
Totality Starts at 11:42 / Lasts 2:22

 45 Casper Mountain County Park
Casper, WY 82601
42.74186 -106.31061
Totality Starts at 11:42 / Lasts 2:27

 46 Goodstein Park
S Walnut St, Casper, WY 82601
42.79787 -106.33785
Totality Starts at 11:42 / Lasts 2:27

 47 Wolf Creek Park
4130 Puma Dr, Casper, WY 82604
42.80996 -106.37028
Totality Starts at 11:42 / Lasts 2:27

 48 Morad Park
2800 SW Wyoming Blvd, Casper, WY
42.82110 -106.37229
Totality Starts at 11:42 / Lasts 2:27

 49 CP HWY 258
2090 SW Wyoming Blvd, Casper, WY
42.83017 -106.36621
Totality Starts at 11:42 / Lasts 2:26

 50 CP HWY 220
2976 CY Ave, Casper, WY 82604
42.82815 -106.35620
Totality Starts at 11:42 / Lasts 2:26

 51 CP HWY 251
Casper Mountain Rd, Casper, WY
42.82267 -106.32891
Totality Starts at 11:42 / Lasts 2:26

Plan to be at your eclipse viewing site at least two hours before totality starts.

52 CP HWY 258
3099 SE Wyoming Blvd, Casper, WY
42.81498 -106.29057
Totality Starts at 11:42 / Lasts 2:27

61 Casper Speedway
1277-1345 Amoco Rd, Casper, WY
42.87681 -106.32008
Totality Starts at 11:42 / Lasts 2:24

53 **Casper City Park**
711 S Center St, Casper, WY 82601
42.84328 -106.32480
Totality Starts at 11:42 / Lasts 2:26

62 **Casper Events Center**
1 Events Dr, Casper, WY 82601
42.86965 -106.33229
Totality Starts at 11:42 / Lasts 2:25

54 **Washington Park**
951 S Jefferson St, Casper, WY 82601
42.84082 -106.31369
Totality Starts at 11:42 / Lasts 2:26

63 **Crossroads Park**
1101 N Poplar St, Casper, WY 82601
42.86292 -106.33098
Totality Starts at 11:42 / Lasts 2:25

55 **Highland Park**
630 S Beverly St, Casper, WY 82601
42.84425 -106.29947
Totality Starts at 11:42 / Lasts 2:26

64 **Mathew Campfield Park**
1219 N Beech St, Casper, WY 82601
42.86297 -106.31970
Totality Starts at 11:42 / Lasts 2:25

56 **Sage Park**
2930 E 15th St, Casper, WY 82609
42.83640 -106.28745
Totality Starts at 11:42 / Lasts 2:26

65 **Riverview Park**
1032 E L St, Casper, WY 82601
42.86297 -106.31407
Totality Starts at 11:42 / Lasts 2:25

57 **Fun Valley Park**
3631 E 21st St, Casper, WY 82609
42.82862 -106.27810
Totality Starts at 11:42 / Lasts 2:26

66 **North Casper Sports Complex**
1700 E K St, Casper, WY 82601
42.86348 -106.30294
Totality Starts at 11:42 / Lasts 2:25

58 **Long Park**
501 Shannon Dr, Casper, WY 82609
42.84623 -106.26994
Totality Starts at 11:42 / Lasts 2:25

67 **Rivers Edge Resort**
6820 Santa Fe Cir, Evansville, WY
42.85829 -106.21378
Totality Starts at 11:42 / Lasts 2:24

59 **Verda James Park**
801 Carriage Ln, Casper, WY 82609
42.84191 -106.26588
Totality Starts at 11:42 / Lasts 2:26

68 **Edness Kimball Wilkins SP**
Evansville, WY 82636
42.85440 -106.17818
Totality Starts at 11:42 / Lasts 2:24

60 **Susie McMurry Park**
5135 E 21st St, Casper, WY 82609
42.82936 -106.25516
Totality Starts at 11:42 / Lasts 2:26

69 CP 6284 Hat 6 Rd,
Casper, WY 82609
42.78623 -106.14850
Totality Starts at 11:42 / Lasts 2:27

Plan to be at your eclipse viewing site at least two hours before totality starts.

Eastern Wyoming

70 Municipal Park
Glenrock, WY 82637
42.86483 -105.86509
Totality Starts at 11:43 / Lasts 2:19

71 Ayres Natural Bridge Park
208 Natural Bridge Rd, Douglas, WY
42.73483 -105.61114
Totality Starts at 11:43 / Lasts 2:25

72 Bartling Park
Douglas, WY 82633
42.74556 -105.37644
Totality Starts at 11:44 / Lasts 2:22

73 Orin Junction Truck Stop
75 US-18 Douglas, WY 82633
42.65821 -105.19049
Totality Starts at 11:44 / Lasts 2:25

74 CP I-25, US-26
Glendo, WY 82213
42.55453 -105.04055
Totality Starts at 11:45 / Lasts 2:28

75 Glendo State Park
397 Glendo Park Rd, Glendo, WY
42.47453 -104.94688
Totality Starts at 11:45 / Lasts 2:29

76 Dwyer Junction Rest Area
Wheatland, WY 82201
42.23462 -105.01741
Totality Starts at 11:45 / Lasts 1:58

77 Guernsey State Park
25 Lake Side Dr, Guernsey, WY
42.28226 -104.76768
Totality Starts at 11:45 / Lasts 2:15

78 CP 1271 Hartville Hwy
Glendo, WY 82213
42.48038 -104.69907
Totality Starts at 11:45 / Lasts 2:29

79 Fort Laramie Historic Site
Gray Rocks Rd, Fort Laramie, WY
42.20359 -104.54517
Totality Starts at 11:46 / Lasts 2:11

80 Chuck Wagon RV Park
310 Pioneer Ct, Fort Laramie, WY
42.21343 -104.52238
Totality Starts at 11:46 / Lasts 2:13

81 CP U.S. 85, Canam Hwy,
Torrington, WY 82240
42.40463 -104.35651
Totality Starts at 11:46 / Lasts 2:29

82 Lingle Town Park
W 4th St, Lingle, WY 82223
42.13754 -104.34670
Totality Starts at 11:46 / Lasts 2:08

83 CP Van Tassell Rd,
Torrington, WY 82240
42.36205 -104.16660
Totality Starts at 11:46 / Lasts 2:29

84 Goshen County Fairgrounds
7078 Fairgrounds Rd, Torrington, WY
42.07207 -104.20019
Totality Starts at 11:46 / Lasts 2:02

85 Jirdon Park
Torrington, WY 82240
42.07034 -104.17543
Totality Starts at 11:46 / Lasts 2:03

86 Torrington Municipal Airport
Torrington, WY 82240
42.06465 -104.15674
Totality Starts at 11:47 / Lasts 2:03

Plan to be at your eclipse viewing site at least two hours before totality starts.

182 Places to See the Total Eclipse in Nebraska

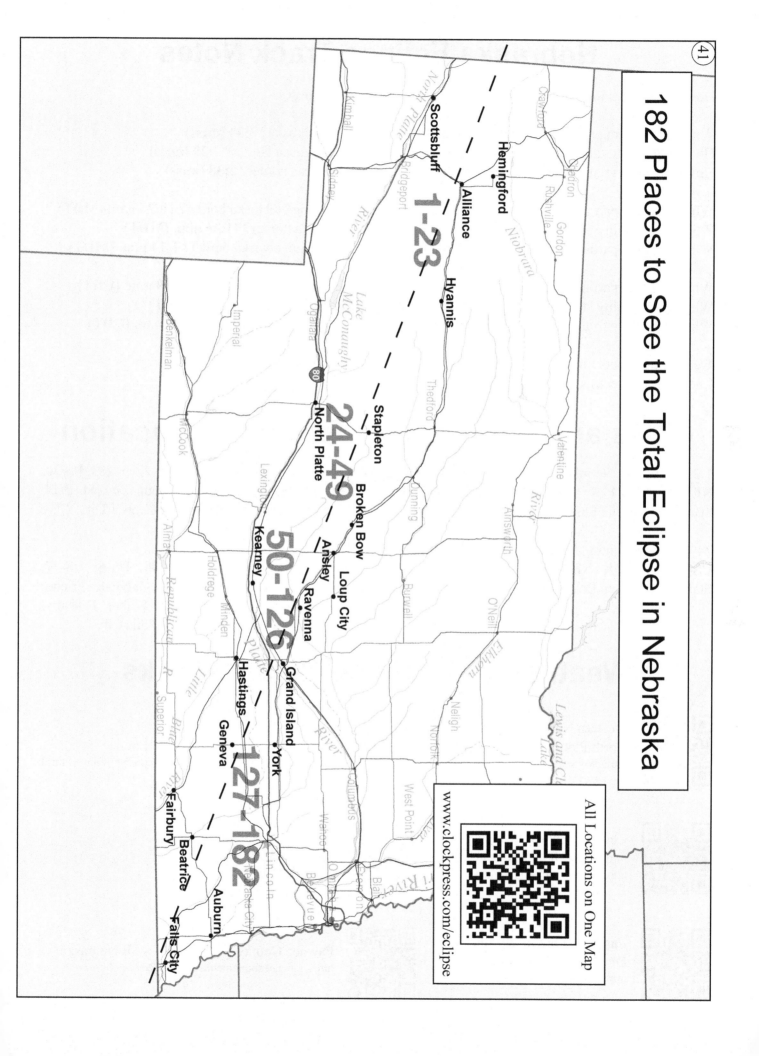

All Locations on One Map

www.clockpress.com/eclipse

Nebraska Eclipse Track Notes

How many miles long is the eclipse centerline in Nebraska? **468 miles**

What is the average duration of totality along the centerline in Nebraska? **2:34 (m:ss)**
How long does totality last on the centerline at the Wyoming/Nebraska border? **2:29 (m:ss)**
How long does totality last on the centerline at the Nebraska/Kansas border? **2:38 (m:ss)**

When does the partial eclipse start on the centerline at the Wyoming/Nebraska border? **10:25 a.m. (MDT)**
When does totality start on the centerline at the Wyoming/Nebraska border? **11:46 a.m. (MDT)**
When does the partial eclipse end on the centerline at the Wyoming/Nebraska border? **1:14 p.m. (MDT)**

When does the partial eclipse start on the centerline at the Nebraska/Kansas border? **11:39 a.m. (CDT)**
When does totality start on the centerline at the Nebraska/Kansas border? **1:04 p.m. (CDT)**
When does the partial eclipse end on the centerline at the Nebraska/Kansas border? **2:32 p.m. (CDT)**

Weather prospects for this event in Nebraska? **Best in western Nebraska.**
Check local forecasts and satellite maps for the best weather information for this eclipse.

Cities and Highways in Totality by Location

1-23: Western Nebraska - **US-26, US-385, NE-2, NE-71, NE-61** - Morrill, Harrison, Mitchell, Scottsbluff, Agate Fossil Beds, Alliance

24-42: West Central Nebraska - **US-83, I-80, NE-2, US-183, NE-21, NE-97, NE-40** - North Platte, Stapleton, Broken Bow, Ansley

43-126: Central Nebraska - **I-80, NE-2, US-281, US-30, US-34, NE-11, NE-14, US-6** - Loup City, St. Paul, Kearney, Grand Island, Hastings, Central City, Clay Center, York

127-179: Eastern Nebraska - **I-80, US-6, US-77, US-136, US-75, US-73, NE-41, NE-4, NE-8** - Friend, Denton, Hickman, Martell, Crete, Wilber, Fairbury, Beatrice, Humboldt, Pawnee City, Falls City

Weather and Eclipse Related Links

Website

Western Nebraska Weather
North Platte, NE NWS Office
http://www.weather.gov/lbf/

Central Plains
U.S. GOES Satellite - Visible Loop
http://www.ssd.noaa.gov/goes/east/cp/h5-mloop-vis.html

Central Nebraska Weather
Hastings, NE NWS Office
http://www.weather.gov/gid/

North Platte Eclipse
http://2017nebraskaeclipse.com

Eastern Nebraska Weather
Omaha, NE NWS Office
http://www.weather.gov/oax/

Pawnee County Eclipse 2017 - SE Nebraska
http://pawneecountynebraska.com/2017-eclipse/

Featured Eclipse Destinations

Agate Fossil Beds National Monument in western Nebraska has over 3 miles of trails, fossil exhibits and the exciting Cook Collection of Native American artifacts. Big skies and two minutes twenty seconds of totality will greet monument visitors on eclipse day. Totality starts at 11:47 a.m. MDT. See Nebraska #4 for location details.

Website

Photos

Homestead National Monument features many outdoor and indoor activities highlighting the Homestead Act of 1862. Visitors to this national monument will see over two and a half minutes of totality on eclipse day. Totality starts at 1:02 p.m. CDT. See Nebraska #147 for location details.

Website

Photos

Carhenge is a replica of Stonehenge made with automobiles. Positioned just north of Alliance, this site is hosting a large eclipse watching event. The city of Alliance will be hosting a three day celebration and solar viewing at Laing Park and the Alliance airport. Totality starts at 11:49 a.m. MDT. See Nebraska #13 for location details.

Website

Photos

Indian Cave State Park covers 3,052 acres along the Missouri river in eastern Nebraska. Hiking, boating, fishing and camping are just a few of the outdoor activities available at this park. On eclipse day, visitors will be treated to well over two minutes of totality. Totality starts at 1:04 p.m. CDT. See Nebraska #178 for location details.

Website

Photos

NE Crossing Points Table

Where the Eclipse Centerline Crosses Highways in Nebraska

M = Mountain Daylight Time   C = Central Daylight Time

Loc#	Hwy	Nearest Mile Marker, Cross Street or Exit, City	TStart	TLasts
5	NE-71	MM 89, Alliance, NE	11:47M	2:30
23	NE-61	MM 140, Hyannis, NE	11:51M	2:32
24	NE-97	MM 33, Tryon, NE	12:53C	2:33
32	NE-40	Rd 797, Arnold, NE	12:54C	2:34
39	NE-21	MM 61, Ash Creek Canyon Rd, Broken Bow, NE	12:55C	2:35
42	US-183	MM 80, Bridge 07981, Mason City, NE	12:56C	2:35
48	NE-2	MM 327, Pine Rd, Ravenna, NE	12:57C	2:35
70	NE-11	W 13th St, Wood River, NE	12:58C	2:36
94	I-80	MM 319, Grand Island, NE	12:58C	2:36
116	NE-14	W 4 Rd, Aurora, NE	12:59C	2:36
124	US-6	MM 255, RD 12, Fairmont, NE	1:00C	2:37
133	NE-41	County Rd 1400, Wilber, NE	1:01C	2:37
150	US-77	MM 29, Dogwood Rd, Pickrell, NE	1:02C	2:37
162	US-136	Mud Creek, S 148th Rd, Filley, NE	1:02C	2:38
166	NE-4	620 Ave, Steinauer, NE	1:03C	2:38
175	NE-8	Bridge 13678, MM 137, Salem, NE	1:04C	2:38

Western Nebraska

M = Mountain Daylight Time
C = Central Daylight Time

1 Morrill Park
Monroe Ave, Morrill, NE 69358
41.96374 -103.92941
Totality Starts at 11:47M / Lasts 1:54

2 Scottsbluff County Fairgrounds
2299 13th St, Mitchell, NE 69357
41.94003 -103.81926
Totality Starts at 11:47M / Lasts 1:54

3 Mitchell City Pool
2044 17th St, Mitchell, NE 69357
41.94479 -103.81537
Totality Starts at 11:47M / Lasts 1:55

4 Agate Fossil Beds National Monument
301 River Rd, Harrison, NE 69346
42.42578 -103.73317
Totality Starts at 11:47M / Lasts 2:20

5 CP 8391 NE-71
Alliance, NE 69301
42.24636 -103.65941
Totality Starts at 11:47M / Lasts 2:30

6 Landers Memorial Soccer Complex
E 42nd St, Scottsbluff, NE 69361
41.89211 -103.65524
Totality Starts at 11:48M / Lasts 1:53

7 Terry Carpenter Park and Recreation Area
Scottsbluff, NE 69361
41.87743 -103.65815
Totality Starts at 11:48M / Lasts 1:50

8 Lakeview Point Campground
Minatare Lake, Minatare, NE 69356
41.92974 -103.50074
Totality Starts at 11:48M / Lasts 2:06

9 Scouts Rest Campground
Minatare Lake, Minatare, NE 69356
41.94626 -103.50400
Totality Starts at 11:48M / Lasts 2:09

10 North Cove Campground
Minatare Lake SRA, Minatare, NE
41.94849 -103.49174
Totality Starts at 11:48M / Lasts 2:10

11 Box Butte Reservoir SRA
Crawford, NE 69339
42.45545 -103.07933
Totality Starts at 11:49M / Lasts 1:50

12 Hemingford City Park
625 Park Ave, Hemingford, NE 69348
42.32038 -103.08049
Totality Starts at 11:48M / Lasts 2:16

13 Carhenge
2151 Co Rd 59, Alliance, NE 69301
42.14219 -102.85819
Totality Starts at 11:49M / Lasts 2:27

14 Laing Park
Alliance, NE 69301
42.11468 -102.87547
Totality Starts at 11:49M / Lasts 2:29

15 Sudman Field
Alliance, NE 69301
42.11148 -102.87174
Totality Starts at 11:49M / Lasts 2:29

16 St. Joseph's Park and Memorial Garden
498 W 11th St, Alliance, NE 69301
42.10621 -102.87595
Totality Starts at 11:49M / Lasts 2:29

17 Hal Murray Softball Complex
1233 4th St, Alliance, NE 69301
42.09908 -102.88732
Totality Starts at 11:49M / Lasts 2:30

Plan to be at your eclipse viewing site at least two hours before totality starts.

Scan for Map

18 Knight Park
Dakota Ave, Alliance, NE 69301
42.10082 -102.88264
Totality Starts at 11:49M / Lasts 2:30

27 Armory Park
N Ash St, North Platte, NE 69101
41.15106 -100.76413
Totality Starts at 12:54C / Lasts 1:52

Scan for Map

19 Central Park
E 10th St, Alliance, NE 69301
42.10421 -102.86809
Totality Starts at 11:49M / Lasts 2:29

28 Cody Park
1401 S Jeffers St, North Platte, NE
41.14785 -100.76060
Totality Starts at 12:54C / Lasts 1:52

20 Bower Park
E 2nd St, Alliance, NE 69301
42.09574 -102.86261
Totality Starts at 11:49M / Lasts 2:30

29 CreekSide Event Center
N Long School Rd, North Platte, NE
41.13785 -100.69325
Totality Starts at 12:54C / Lasts 1:53

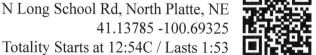

21 CP Alliance Municipal Airport
5631 Sarpy Rd, Alliance, NE 69301
42.04228 -102.79477
Totality Starts at 11:49M / Lasts 2:31

30 Stapleton City Park
NE-92, Stapleton, NE 69163
41.48058 -100.51457
Totality Starts at 12:53C / Lasts 2:33

22 Frye Lake
NE-61, Hyannis, NE 69350
42.01631 -101.75319
Totality Starts at 11:51M / Lasts 2:15

Central and Eastern Nebraska Start Times are
Central Daylight Time

Central Nebraska

23 CP NE-61
Hyannis, NE 69350
41.77919 -101.73104
Totality Starts at 11:51M / Lasts 2:32

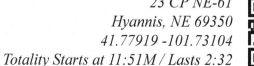

31 Arnold State Recreation Area
Arnold, NE 69120
41.41340 -100.19971
Totality Starts at 12:54 / Lasts 2:33

24 CP Glenn Miller Memorial Hwy
Tryon, NE 69167
41.58085 -100.96294
Totality Starts at 12:53C / Lasts 2:33

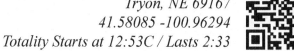

32 CP NE-40
Arnold, NE 69120
41.37649 -100.19882
Totality Starts at 12:54 / Lasts 2:34

25 Lincoln County Fairgrounds
5015 Rodeo Rd, North Platte, NE 69101
41.16129 -100.81518
Totality Starts at 12:53C / Lasts 1:51

33 Morgan Park Campground
Callaway, NE 68825
41.29249 -99.92258
Totality Starts at 12:55 / Lasts 2:35

26 Buffalo Bill State Recreation Area
Scouts Rest Ranch Rd, North Platte, NE
41.16255 -100.79272
Totality Starts at 12:53C / Lasts 1:53

34 Victoria Springs SRA
Anselmo, NE 68813
41.61000 -99.75024
Totality Starts at 12:55 / Lasts 1:52

Plan to be at your eclipse viewing site at least two hours before totality starts.

Scan for Map

35 Indian Hills Park
N 17th Ave, Broken Bow, NE 68822
41.41237 -99.64978
Totality Starts at 12:55 / Lasts 2:22

36 North Park
N J St, Broken Bow, NE 68822
41.41225 -99.64214
Totality Starts at 12:55 / Lasts 2:22

37 Melham Park
Laural Dr, Broken Bow, NE 68822
41.41365 -99.63496
Totality Starts at 12:55 / Lasts 2:22

38 South Park
S 5th Ave, Broken Bow, NE 68822
41.39556 -99.63519
Totality Starts at 12:55 / Lasts 2:24

39 CP NE-21
Broken Bow, NE 68822
41.23345 -99.67934
Totality Starts at 12:55 / Lasts 2:35

40 Oconto City Park
Oconto, NE 68860
41.14192 -99.76319
Totality Starts at 12:55 / Lasts 2:33

41 Ansley City Park
Ansley, NE 68814
41.29061 -99.38835
Totality Starts at 12:56 / Lasts 2:27

42 CP US-183
Mason City, NE 68855
41.14916 -99.37857
Totality Starts at 12:56 / Lasts 2:35

43 Arcadia Wayside Park
Arcadia, NE 68815
41.42379 -99.12459
Totality Starts at 12:57 / Lasts 1:57

Scan for Map

44 Loup City Recreation Area
Loup City, NE 68853
41.27700 -98.99169
Totality Starts at 12:57 / Lasts 2:17

45 Jenners Park
N 1st Ave, Loup City, NE 68853
41.27464 -98.95748
Totality Starts at 12:57 / Lasts 2:16

46 Redwood Area Campground
Sherman Reservoir, Loup City, NE
41.29187 -98.89852
Totality Starts at 12:57 / Lasts 2:10

47 Sherman Reservoir SRA
79025 Sherman Dam Rd
Loup City, NE 68853
41.32865 -98.89090
Totality Starts at 12:57 / Lasts 2:03

48 CP 42177 NE-2
Ravenna, NE 68869
41.01786 -98.91788
Totality Starts at 12:57 / Lasts 2:35

49 Ravenna Lake SRA
Ravenna, NE 68869
41.01907 -98.89240
Totality Starts at 12:57 / Lasts 2:35

50 Cottonmill Park
2795 Cottonmill Ave, Kearney, NE
40.70322 -99.14782
Totality Starts at 12:57 / Lasts 1:54

51 Cottonmill Lake SRA
Kearney, NE 68845
40.70543 -99.14219
Totality Starts at 12:57 / Lasts 1:54

52 West Lincolnway Park
30th Ave, Kearney, NE 68845
40.69733 -99.12173
Totality Starts at 12:57 / Lasts 1:54

Plan to be at your eclipse viewing site at least two hours before totality starts.

Scan for Map

Scan for Map

53 Yanney Park
Kearney, NE 68845
40.68205 -99.10746
Totality Starts at 12:57 / Lasts 1:51

62 Great Platte River Road Archway
3060 E 1st St, Kearney, NE 68847
40.67054 -99.03779
Totality Starts at 12:57 / Lasts 1:53

54 Fountain Hills Park
W 48th St, Kearney, NE 68845
40.71996 -99.09682
Totality Starts at 12:57 / Lasts 2:00

63 Fort Kearny SRA
Gibbon, NE 68840
40.65530 -98.99465
Totality Starts at 12:57 / Lasts 1:53

55 Memorial Diamond Park
Kearney, NE 68845
40.70980 -99.09222
Totality Starts at 12:57 / Lasts 1:58

64 Kearney Raceway Park
4860 Imperial Ave, Kearney, NE
40.72814 -98.98936
Totality Starts at 12:57 / Lasts 2:07

56 Pioneer Park
W 21st St, Kearney, NE 68845
40.69643 -99.09201
Totality Starts at 12:57 / Lasts 1:55

65 Davis Park
River St, Gibbon, NE 68840
40.75273 -98.84767
Totality Starts at 12:57 / Lasts 2:19

57 Centenial Park
Kearney, NE 68845
40.68622 -99.09015
Totality Starts at 12:57 / Lasts 1:54

66 Windmill State Recreation Area
Gibbon, NE 68840
40.70673 -98.84089
Totality Starts at 12:57 / Lasts 2:11

58 Dryden Park
Kearney, NE 68847
40.70627 -99.07203
Totality Starts at 12:57 / Lasts 1:59

67 Forest Park
N C St, Shelton, NE 68876
40.78400 -98.73325
Totality Starts at 12:57 / Lasts 2:28

59 Harvey Park
E 44th St, Kearney, NE 68847
40.71675 -99.06704
Totality Starts at 12:57 / Lasts 2:01

68 Memorial Park
E 4th St, Shelton, NE 68876
40.77629 -98.72871
Totality Starts at 12:57 / Lasts 2:27

60 Buffalo County Fairgrounds
3807 Avenue N, Kearney, NE 68847
40.71118 -99.06201
Totality Starts at 12:57 / Lasts 2:00

69 War Axe SRA
I-80, Shelton, NE 68876
40.72451 -98.73523
Totality Starts at 12:58 / Lasts 2:19

61 Nebraska Firefighters Museum
2434 E 1st St, Kearney, NE 68847
40.67116 -99.04519
Totality Starts at 12:57 / Lasts 1:53

70 CP NE-11
Wood River, NE 68883
40.92803 -98.60802
Totality Starts at 12:58 / Lasts 2:36

Plan to be at your eclipse viewing site at least two hours before totality starts.

Scan for Map

71 Library Park
West St, Wood River, NE 68883
40.82130 -98.60241
Totality Starts at 12:58 / Lasts 2:35

72 Memorial Park
E 13th St, Wood River, NE 68883
40.82536 -98.59454
Totality Starts at 12:58 / Lasts 2:35

73 Cheyenne State Recreation Area
I-80, Wood River, NE 68883
40.76406 -98.59223
Totality Starts at 12:58 / Lasts 2:30

74 Cheyenne State Recreation Area
Wood River, NE 68883
40.76504 -98.59021
Totality Starts at 12:58 / Lasts 2:31

75 Dannebrog Park
Dannebrog, NE 68831
41.11752 -98.54513 Totality Starts at
12:58 / Lasts 2:22

76 Grover Cleveland Ball Field
St Paul, NE 68873
41.21181 -98.46861
Totality Starts at 12:58 / Lasts 2:03

77 St Paul City Park
Jay St, St Paul, NE 68873
41.21494 -98.45511
Totality Starts at 12:58 / Lasts 2:01

78 Eagle Scout Park
Grand Island, NE 68803
40.95952 -98.36102
Totality Starts at 12:58 / Lasts 2:31

79 George Park
Grand Island, NE 68803
40.94344 -98.40426
Totality Starts at 12:58 / Lasts 2:32

Scan for Map

80 Veterans Park
Grand Island, NE 68803
40.94761 -98.36034
Totality Starts at 12:58 / Lasts 2:31

81 Sprague Park
Grand Island, NE 68801
40.94076 -98.32683
Totality Starts at 12:58 / Lasts 2:31

82 Ryder Park
W North Front St, Grand Island, NE
40.91935 -98.37120
Totality Starts at 12:58 / Lasts 2:33

83 Buechler Park
W Division St, Grand Island, NE 68803
40.91340 -98.36384
Totality Starts at 12:58 / Lasts 2:33

84 Pier Park
500 S Oak St, Grand Island, NE 68801
40.91766 -98.33510
Totality Starts at 12:58 / Lasts 2:33

85 Augustine Park
Garland St, Grand Island, NE 68803
40.90932 -98.36442
Totality Starts at 12:58 / Lasts 2:34

86 Fonner Park
700 E Stolley Park Rd
Grand Island, NE 68802
40.90624 -98.32685
Totality Starts at 12:58 / Lasts 2:33

87 Stolley State Park
2822 W Stolley Park Rd
Grand Island, NE 68801
40.90049 -98.36174
Totality Starts at 12:58 / Lasts 2:34

88 Ray Aquatic Park
Grand Island, NE 68801
40.88970 -98.35257
Totality Starts at 12:58 / Lasts 2:35

Plan to be at your eclipse viewing site at least two hours before totality starts.

89 Museum of the Prairie Pioneer
3133 US-34 Grand Island, NE 68801
40.88480 -98.37317
Totality Starts at 12:58 / Lasts 2:35

98 American Legion Park
N Minnesota Ave, Hastings, NE 68901
40.60679 -98.38378
Totality Starts at 12:58 / Lasts 2:17

90 George Clayton Hall County Park
Grand Island, NE 68801
40.87663 -98.37077
Totality Starts at 12:58 / Lasts 2:35

99 Libbs Park
N Baltimore Ave, Hastings, NE 68901
40.60328 -98.40264
Totality Starts at 12:58 / Lasts 2:16

91 Mormon Island SRA
7425 US-281 Doniphan, NE 68832
40.82570 -98.37034
Totality Starts at 12:58 / Lasts 2:37

100 Heartwell Park
Lakeside Dr, Hastings, NE 68901
40.59505 -98.37835
Totality Starts at 12:58 / Lasts 2:16

92 Grand Island Historical Marker
Doniphan, NE 68832
40.82289 -98.32556
Totality Starts at 12:58 / Lasts 2:37

101 Oswego Park
W 7th St, Hastings, NE 68901
40.58933 -98.40795
Totality Starts at 12:58 / Lasts 2:13

93 Grand Island KOA
904 South B Road, Doniphan, NE 68832
40.81580 -98.26364
Totality Starts at 12:58 / Lasts 2:36

102 Brickyard Park
Hastings, NE 68901
40.57769 -98.40953
Totality Starts at 12:58 / Lasts 2:11

94 CP I-80
Grand Island NE 68832
40.82100 -98.24404
Totality Starts at 12:58 / Lasts 2:36

103 Adams County Fairgrounds
947 S Baltimore Ave, Hastings, NE
40.57220 -98.40196
Totality Starts at 12:58 / Lasts 2:10

95 Prairie Lake SRA
Hastings, NE 68901
40.53469 -98.48973
Totality Starts at 12:58 / Lasts 1:59

104 Duncan Field
E South St, Hastings, NE 68901
40.58356 -98.37461
Totality Starts at 12:58 / Lasts 2:14

96 Prairie Ridge Park
W 42nd St, Hastings, NE 68901
40.62539 -98.37768
Totality Starts at 12:58 / Lasts 2:21

105 Fergus Park
Hastings, NE 68901
40.57637 -98.37269
Totality Starts at 12:58 / Lasts 2:12

97 Hastings KOA Campground
Hastings, NE 68901
40.61124 -98.37699
Totality Starts at 12:58 / Lasts 2:18

106 DLD State Wayside Campground
U.S-6, Hastings, NE 68901
40.58206 -98.28970
Totality Starts at 12:59 / Lasts 2:18

Plan to be at your eclipse viewing site at least two hours before totality starts.

107 Fairfield City Park
299 E 8th St, Fairfield, NE 68938
40.43507 -98.10371
Totality Starts at 12:59 / Lasts 2:02

108 Clay Center Park
W Johnson St, Clay Center, NE 68933
40.52461 -98.05775
Totality Starts at 12:59 / Lasts 2:20

109 Clay County Fairgrounds
701 N Martin Ave, Clay Center, NE
40.52743 -98.05906
Totality Starts at 12:59 / Lasts 2:20

110 Harvard City Park
E 9th St, Harvard, NE 68944
40.62417 -98.09498
Totality Starts at 12:59 / Lasts 2:32

111 Bader Memorial Park
898 Bader Park Rd, Chapman, NE
40.98608 -98.15036
Totality Starts at 12:58 / Lasts 2:24

112 South Park
2502 16th Ave, Central City, NE 68826
41.10662 -98.00077
Totality Starts at 12:59 / Lasts 1:55

113 Hord Lake SRA
Central City, NE 68826
41.10559 -97.95615
Totality Starts at 12:59 / Lasts 1:50

114 Streeter Park
Aurora, NE 68818
40.87461 -98.00311
Totality Starts at 12:59 / Lasts 2:30

115 Plainsman Museum
210 16th St, Aurora, NE 68818
40.85904 -97.99668
Totality Starts at 12:59 / Lasts 2:31

116 CP 476 NE-14
Aurora, NE 68818
40.74741 -97.99696
Totality Starts at 12:59 / Lasts 2:36

117 Sutton City Park
Sutton, NE 68979
40.61094 -97.85827
Totality Starts at 12:59 / Lasts 2:36

Eastern Nebraska

118 Miller Park
York, NE 68467
40.87815 -97.58537
Totality Starts at 1:00 / Lasts 2:16

119 Duke Park
York, NE 68467
40.87601 -97.58039
Totality Starts at 1:00 / Lasts 2:16

120 Harrison Park
York, NE 68467
40.86280 -97.60124
Totality Starts at 1:00 / Lasts 2:19

121 Foster Park Arboretum
York, NE 68467
40.86412 -97.58666
Totality Starts at 1:00 / Lasts 2:18

122 Beaver Creek Park
York, NE 68467
40.85978 -97.59039
Totality Starts at 1:00 / Lasts 2:19

123 Northside Park
6th St, McCool Junction, NE 68401
40.74777 -97.59217
Totality Starts at 1:00 / Lasts 2:30

Plan to be at your eclipse viewing site at least two hours before totality starts.

Scan for Map

124 CP U.S. 6
Fairmont, NE 68354
40.63469 -97.62348
Totality Starts at 1:00 / Lasts 2:37

133 CP 1428 NE-41
Wilber, NE 68465
40.48026 -97.12092
Totality Starts at 1:01 / Lasts 2:37
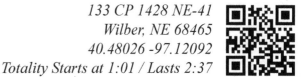

125 Crosstrails Wayside Park
Fairmont, NE 68354
40.63534 -97.59432
Totality Starts at 1:00 / Lasts 2:36

134 Western City Park
Western, NE 68464
40.39308 -97.19667
Totality Starts at 1:01 / Lasts 2:36

126 Geneva City Park
Geneva, NE 68361
40.52255 -97.59346
Totality Starts at 1:00 / Lasts 2:36

135 Fairbury City Park
Park Rd, Fairbury, NE 68352
40.13908 -97.18929
Totality Starts at 1:01 / Lasts 2:00

127 Conns Park
Utica, NE 68456
40.89069 -97.34849
Totality Starts at 1:00 / Lasts 2:01

136 Swanton City Park
Swanton, NE 68445
40.37954 -97.08003
Totality Starts at 1:01 / Lasts 2:37

128 Burley Park
Friend, NE 68359
40.64229 -97.28552
Totality Starts at 1:00 / Lasts 2:31

137 Rock Creek Station
State Historical Park
57426 710th Rd, Fairbury, NE 68352
40.11320 -97.05920
Totality Starts at 1:01 / Lasts 2:02

129 Blizzard 1888 Historical Marker
Milligan, NE 68406
40.52631 -97.38571
Totality Starts at 1:00 / Lasts 2:37

138 Tuxedo Park
Crete, NE 68333
40.63110 -96.97280
Totality Starts at 1:01 / Lasts 2:24

130 Alexandria SRA
Alexandria, NE 68303
40.23409 -97.33732
Totality Starts at 1:01 / Lasts 2:08

139 Little League Park
Crete, NE 68333
40.62085 -96.95904
Totality Starts at 1:01 / Lasts 2:25

131 Blue River SRA
Milford, NE 68405
40.70631 -97.12023
Totality Starts at 1:01 / Lasts 2:21

140 Wildwood Park
Crete, NE 68333
40.61646 -96.95394
Totality Starts at 1:01 / Lasts 2:25

132 Dorchester City Park
Dorchester, NE 68343
40.64738 -97.11704
Totality Starts at 1:01 / Lasts 2:27

141 Wilber City Park
Wilber, NE 68465
40.48192 -96.96839
Totality Starts at 1:01 / Lasts 2:35

Plan to be at your eclipse viewing site at least two hours before totality starts.

Scan for Map

 142 Legion Memorial Park
Wilber, NE 68465
40.47451 -96.96289
Totality Starts at 1:01 / Lasts 2:35

 143 Memorial Park
De Witt, NE 68341
40.39557 -96.91948
Totality Starts at 1:01 / Lasts 2:38

 144 Diller Campground
Castor St, Diller, NE 68342
40.10982 -96.93447
Totality Starts at 1:02 / Lasts 2:09

 145 Conestoga Lake SRA
Denton, NE 68339
40.77065 -96.85152
Totality Starts at 1:01 / Lasts 1:51

 146 Olive Creek Lake SRA
Hallam, NE 68368
40.58008 -96.84755
Totality Starts at 1:01 / Lasts 2:26

 147 Homestead National Monument
8523 NE-4, Beatrice, NE 68310
40.29159 -96.83574
Totality Starts at 1:02 / Lasts 2:37

 148 Bluestem Lake SRA
Martell, NE 68404
40.63918 -96.79560
Totality Starts at 1:01 / Lasts 2:16

 149 Cortland Community Park
Lincoln St, Cortland, NE 68331
40.50420 -96.70755
Totality Starts at 1:02 / Lasts 2:29

 150 CP 17644 Homestead Expy
Pickrell, NE 68422
40.36176 -96.74227
Totality Starts at 1:02 / Lasts 2:37

Scan for Map

 151 Beatrice Municipal Airport
Beatrice, NE 68310
40.31002 -96.75251
Totality Starts at 1:02 / Lasts 2:38

 152 Astro Park
Beatrice, NE 68310
40.27772 -96.72885
Totality Starts at 1:02 / Lasts 2:37

 153 Nichols Park
US Hwy 136, Beatrice, NE 68310
40.26695 -96.75973
Totality Starts at 1:02 / Lasts 2:37

 154 Chautauqua Park Campground
Beatrice, NE 68310
40.25377 -96.73963
Totality Starts at 1:02 / Lasts 2:36

 155 Arbor Park
N 7th St, Wymore, NE 68466
40.13082 -96.66448
Totality Starts at 1:02 / Lasts 2:27

 156 McCandles Park
W G St, Wymore, NE 68466
40.12235 -96.67112
Totality Starts at 1:02 / Lasts 2:25

 157 Hickman City Park
Hickman, NE 68372
40.61877 -96.63459
Totality Starts at 1:02 / Lasts 2:11

 158 Prairie Park
441 Pioneer Ct, Hickman, NE 68372
40.62293 -96.62737
Totality Starts at 1:02 / Lasts 2:10

 159 Wagon Train Lake SRA
3019 Apple St, Lincoln, NE 68503
40.62654 -96.58127
Totality Starts at 1:02 / Lasts 2:07

Plan to be at your eclipse viewing site at least two hours before totality starts.

Scan for Map

Scan for Map

160 Stagecoach Lake SRA
Hickman, NE 68372
40.59910 -96.63720
Totality Starts at 1:02 / Lasts 2:14

169 Sportsman's Park
Washington, NE 68378
40.41056 -96.00518
Totality Starts at 1:03 / Lasts 2:13

161 Rockford Lake SRA
Beatrice, NE 68310
40.22670 -96.58276
Totality Starts at 1:02 / Lasts 2:37

170 Kirkman's Cove Recreation Area
Humboldt, NE 68376
40.18091 -95.99203
Totality Starts at 1:03 / Lasts 2:35

162 CP 14927 US-136
Filley, NE 68357
40.29088 -96.51856
Totality Starts at 1:02 / Lasts 2:38

171 Town Square Park
Humboldt, NE 68376
40.16332 -95.94719
Totality Starts at 1:03 / Lasts 2:35

163 Adams Community Center
549 5th St, Adams, NE 68301
40.45699 -96.50680
Totality Starts at 1:02 / Lasts 2:28

172 Humboldt Lake Park
Humboldt, NE 68376
40.16003 -95.94792
Totality Starts at 1:03 / Lasts 2:36

164 Sterling Park
300 Lincoln St, Sterling, NE 68443
40.46087 -96.37466
Totality Starts at 1:02 / Lasts 2:23

173 Auburn City Recreation Complex
409 O St, Auburn, NE 68305
40.39936 -95.84826
Totality Starts at 1:04 / Lasts 2:06

165 Tecumseh City Park
Clay St, Tecumseh, NE 68450
40.36689 -96.18881
Totality Starts at 1:03 / Lasts 2:27

174 Legion Memorial Park
Auburn, NE 68305
40.39462 -95.83773
Totality Starts at 1:04 / Lasts 2:06

166 CP NE-4
Steinauer, NE 68441
40.18955 -96.20204
Totality Starts at 1:03 / Lasts 2:38

175 CP 706 Rd, NE-8
Salem, NE 68433
40.05853 -95.79869
Totality Starts at 1:04 / Lasts 2:38

167 City Pond Park
Pawnee City, NE 68420
40.10551 -96.16226
Totality Starts at 1:03 / Lasts 2:38

176 Verdon Lake SRA
Verdon, NE 68457
40.14558 -95.73557
Totality Starts at 1:04 / Lasts 2:33

168 South Park
Pawnee City, NE 68420
40.10437 -96.15398
Totality Starts at 1:03 / Lasts 2:38

177 Lewis and Clark Campsite
Brownville, NE 68321
40.39300 -95.65270
Totality Starts at 1:04 / Lasts 1:50

Plan to be at your eclipse viewing site at least two hours before totality starts.

Scan for Map

178 Indian Cave State Park
65296 720 Rd, Shubert, NE 68437
40.26197 -95.57086
Totality Starts at 1:04 / Lasts 2:14

179 Stanton Lake Park
Falls City, NE 68355
40.07034 -95.61282
Totality Starts at 1:04 / Lasts 2:35

180 Water Tower Mini Park
Falls City, NE 68355
40.07046 -95.60816
Totality Starts at 1:04 / Lasts 2:35

181 Grandview Park
Falls City, NE 68355
40.05612 -95.59894
Totality Starts at 1:04 / Lasts 2:35

182 Piersons Point
Falls City, NE 68355
40.03129 -95.56859
Totality Starts at 1:04 / Lasts 2:36

Plan to be at your eclipse viewing site at least two hours before totality starts.

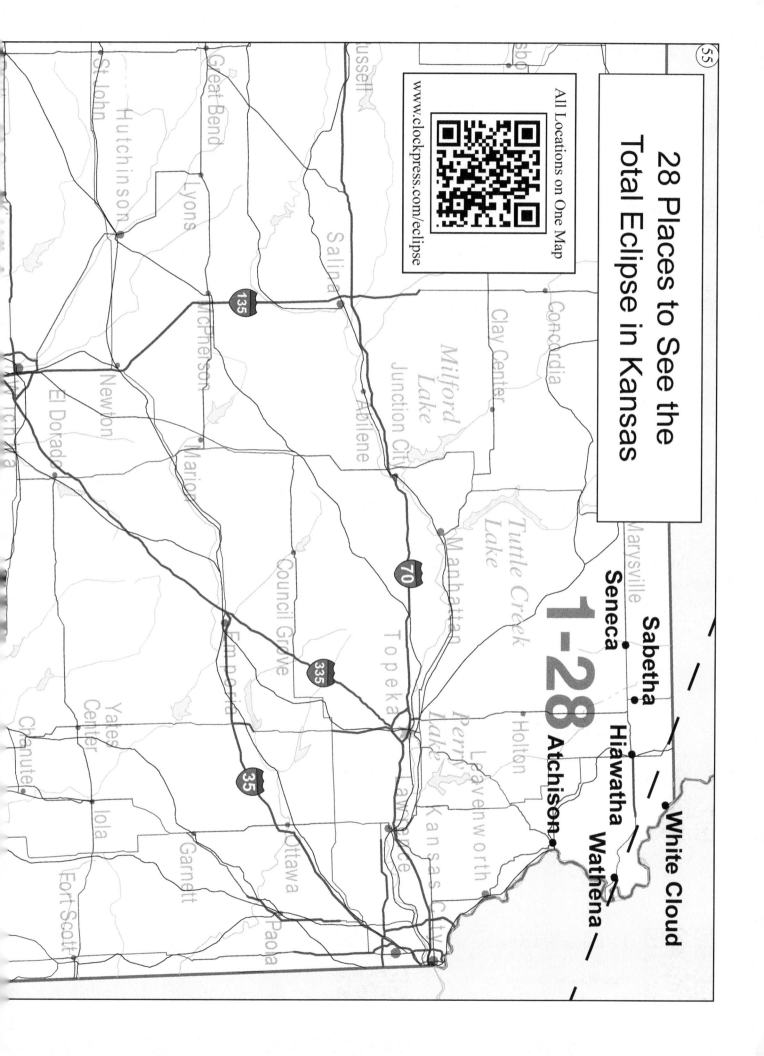

28 Places to See the
Total Eclipse in Kansas

All Locations on One Map

www.clockpress.com/eclipse

White Cloud

Sabetha

Seneca

Hiawatha

Wathena

1-28 Atchison

Northeast Kansas Eclipse Track Notes

How many miles long is the eclipse centerline in Kansas? **44 miles**

What is the average duration of totality along the centerline in Kansas? **2:38 (m:ss)**
How long does totality last on the centerline at the Nebraska/Kansas border? **2:38 (m:ss)**
How long does totality last on the centerline at the Kansas/Missouri border? **2:39 (m:ss)**

When does the partial eclipse start on the centerline at the Nebraska/Kansas border? **11:39 a.m. (CDT)**
When does totality start on the centerline at the Nebraska/Kansas border? **1:04 p.m. (CDT)**
When does the partial eclipse end on the centerline at the Nebraska/Kansas border? **2:32 p.m. (CDT)**

When does the partial eclipse start on the centerline at the Kansas/Missouri border? **11:40 a.m. (CDT)**
When does totality start on the centerline at the Kansas/Missouri border? **1:06 p.m. (CDT)**
When does the partial eclipse end on the centerline at the Kansas/Missouri border? **2:34 p.m. (CDT)**

Check local forecasts and satellite maps for the best weather information for this eclipse.

Cities and Highways in Totality by Location

1-28: Northeast Kansas - **KS-63, US-75, US-73, US-159, US-36, KS-7** - Seneca, Sabetha, Hiawatha, Horton, Effingham, White Cloud, Highland, Atchison, Troy, Wathena, Elwood

Weather and Eclipse Related Links

Website

Northeast Kansas Weather
Topeka, KS NWS Office
http://www.weather.gov/top/

Central Plains
U.S. GOES Satellite - Visible Loop
http://www.ssd.noaa.gov/goes/east/cp/h5-mloop-vis.html

NE Kansas/NW Missouri Weather
Pleasant Hill, MO NWS Office
http://www.weather.gov/eax/

City of Hiawatha, Kansas
https://www.cityofhiawatha.org

Featured Eclipse Destinations

Davis Memorial at Mt. Hope Cemetery in Hiawatha is a popular tourist destination and features a massive memorial filled with statues and a mysterious backstory. Two minutes and thirty seven seconds of totality will darken the skies here on eclipse day. Totality starts at 1:05 p.m. CDT. See Kansas #10 for location details.

Website

Photos

Sparks Flea Market Grounds located about ten miles south of White Cloud on KS-7 is exactly on the centerline of the path of totality. This site hosts a large flea market twice a year, usually in May and around Labor Day. Totality starts at 1:05 p.m. CDT and lasts 2:39. See Kansas #19 for location details.

Website

Photos

Four State Lookout in White Cloud sits high atop a bluff overlooking the Missouri river valley. Iowa, Kansas Nebraska and Missouri can be seen from this spot. A magnificent view of the eastern horizon during two minutes and thirty four seconds of totality is your reward for finding this little gem. Totality starts at 1:05 p.m. CDT. See Kansas #17 for location details.

Website

Photos

Atchison Riverfront Park in downtown Atchison features a veterans memorial and walking trail. Over two minutes of totality will be visible from this scenic attraction. Totality starts at 1:06 p.m. CDT. See Kansas #23 for location details.

Website

Photos

KS Crossing Points Table

Where the Eclipse Centerline Crosses Highways in Kansas					
				Start Times = Central Daylight Time	
Loc#	Hwy	Nearest Mile Marker, Cross Street or Exit, City		TStart	TLasts
7	US-73	.5 mile south of 702 Rd, Hiawatha, KS		1:04	2:38
19	KS-7	240th Rd, Sparks Flea Market, Highland, KS		1:05	2:39
28	US-36	Exit 20, Roseport Rd, Wathena, KS		1:06	2:39

Northeast Kansas

Start Times = Central Daylight Time

1 Stallbaumer RV Park
1701 North St, Seneca, KS 66538
39.84244 -96.07789
Totality Starts at 1:04 / Lasts 2:11

9 Bruning Park
2100 Apache St, Hiawatha, KS 66434
39.85239 -95.55988
Totality Starts at 1:04 / Lasts 2:36

2 Seneca City Park
Seneca, KS 66538
39.83783 -96.07023
Totality Starts at 1:04 / Lasts 2:10

10 Davis Memorial
Mount Hope Cemetery, Hiawatha, KS
39.84925 -95.51591
Totality Starts at 1:05 / Lasts 2:37

3 Nemaha County Wildlife Area
Seneca, KS 66538
39.77491 -96.03332
Totality Starts at 1:04 / Lasts 2:01

11 Hiawatha City Lake
1499 220th St, Hiawatha, KS 66434
39.82682 -95.52779
Totality Starts at 1:05 / Lasts 2:35

4 Sabetha City Lake
N Lake Rd, Sabetha, KS 66534
39.90906 -95.89940
Totality Starts at 1:04 / Lasts 2:31

12 CP 2998 Prairie Rd
Hiawatha, KS 66434
39.93823 -95.43396
Totality Starts at 1:05 / Lasts 2:38

5 6th Street Park
Sabetha KS 66534
39.90583 -95.79659
Totality Starts at 1:04 / Lasts 2:34

13 Brown State Fishing Lake
Robinson, KS 66532
39.84907 -95.37195
Totality Starts at 1:05 / Lasts 2:38

6 Sabetha Pony Creek Lake
Acorn Rd, US-75, Sabetha, KS 66534
39.94699 -95.78208
Totality Starts at 1:04 / Lasts 2:37

14 Mission Lake
Wilson Dr, Horton, KS 66439
39.67299 -95.51831
Totality Starts at 1:05 / Lasts 2:14

7 CP US-159
Hiawatha, KS 66434
39.99178 -95.59569
Totality Starts at 1:04 / Lasts 2:38

15 Atchison County Park
Horton, KS 66439
39.63599 -95.44572
Totality Starts at 1:05 / Lasts 2:11

8 CP 3192 Longspur Rd
Hiawatha, KS 66434
39.96919 -95.52734
Totality Starts at 1:04 / Lasts 2:38

16 Effingham City Park
305 Main St, Effingham, KS 66023
39.52082 -95.40068
Totality Starts at 1:05 / Lasts 1:52

17 Four State Lookout
3rd St, White Cloud, KS 66094
39.97977 -95.29937
Totality Starts at 1:05 / Lasts 2:34

Plan to be at your eclipse viewing site at least two hours before totality starts.

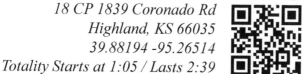

18 CP 1839 Coronado Rd
Highland, KS 66035
39.88194 -95.26514
Totality Starts at 1:05 / Lasts 2:39

Scan for Map

19 CP **Sparks Flea Market Grounds**
KS-7 at 240th Rd, Highland, KS 66035
39.85754 -95.19237
Totality Starts at 1:05 / Lasts 2:39

20 Atchison State Fishing Lake
322nd Rd, Atchison, KS 66002
39.63269 -95.17739
Totality Starts at 1:05 / Lasts 2:26

21 Amelia Earhart Airport
286th Rd, Atchison, KS 66002
39.56914 -95.17867
Totality Starts at 1:06 / Lasts 2:16

22 Reisner Park
798 N 10th St, Atchison, KS 66002
39.56904 -95.12648
Totality Starts at 1:06 / Lasts 2:19

23 Atchison Riverfront Park
136 Commercial St, Atchison, KS 66002
39.56183 -95.11391
Totality Starts at 1:06 / Lasts 2:18

24 Jackson Park
1598 S 6th St, Atchison, KS 66002
39.54422 -95.11781
Totality Starts at 1:06 / Lasts 2:15

25 Doniphan County Fairgrounds
100 N Boder St, Troy, KS 66087
39.78850 -95.09819
Totality Starts at 1:05 / Lasts 2:39

26 CP 1227 Saratoga Rd
Wathena, KS 66090
39.78733 -94.98364
Totality Starts at 1:06 / Lasts 2:39

27 Wathena City Park
304 E St, Joseph St, Wathena, KS 66090
39.76042 -94.94030
Totality Starts at 1:06 / Lasts 2:39

Scan for Map

28 CP 399 Roseport Rd
Wathena, KS 66090
39.75001 -94.87345
Totality Starts at 1:06 / Lasts 2:39

Plan to be at your eclipse viewing site at least two hours before totality starts.

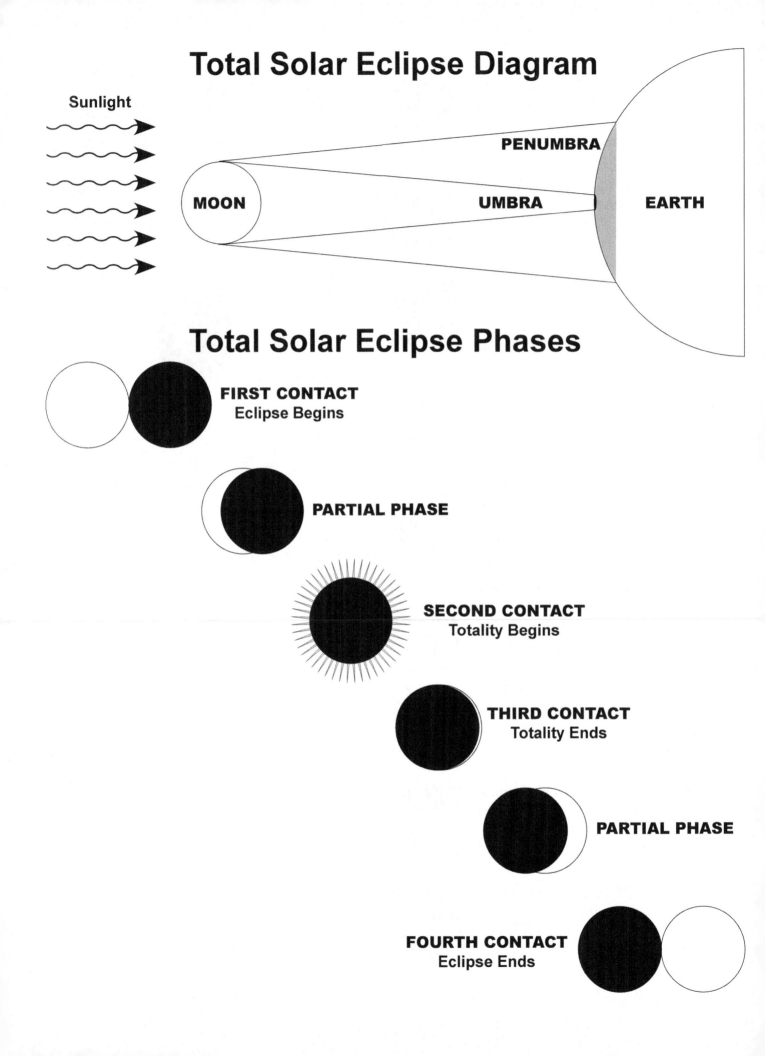

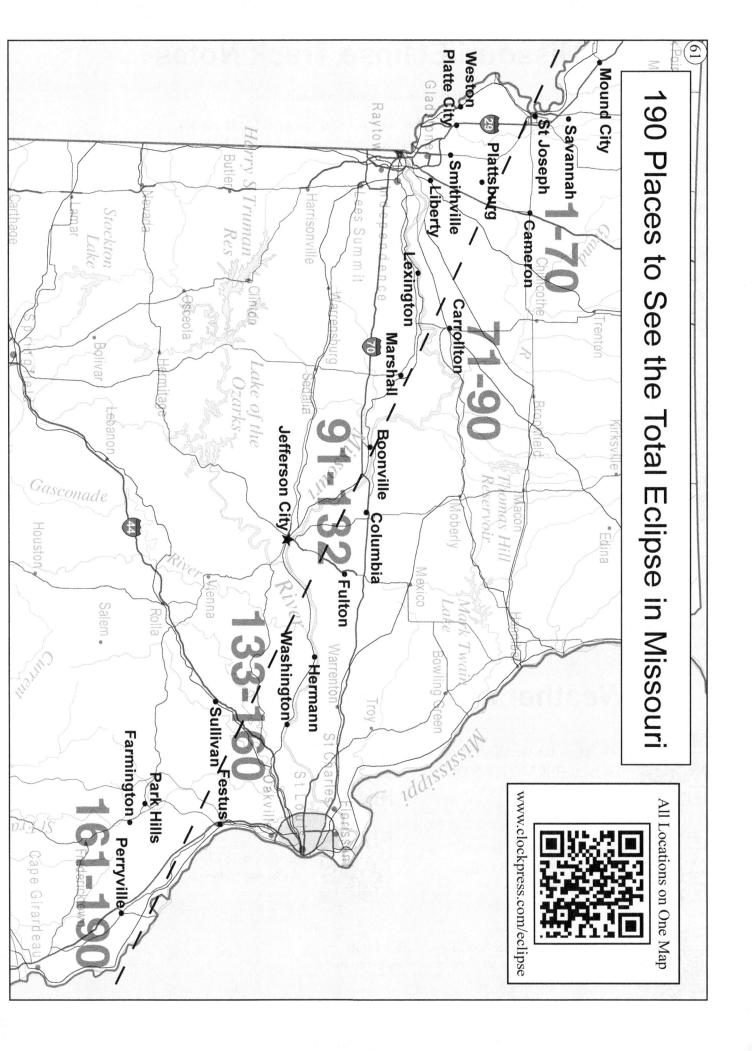

190 Places to See the Total Eclipse in Missouri

All Locations on One Map

www.clockpress.com/eclipse

Missouri Eclipse Track Notes

How many miles long is the eclipse centerline in Missouri? **300 miles**

What is the average duration of totality along the centerline in Missouri? **2:40 (m:ss)**
How long does totality last on the centerline at the Kansas/Missouri border? **2:39 (m:ss)**
How long does totality last on the centerline at the Missouri/Illinois border? **2:41 (m:ss)**

When does the partial eclipse start on the centerline at the Kansas/Missouri border? **11:40 a.m. (CDT)**
When does totality start on the centerline at the Kansas/Missouri border? **1:06 p.m. (CDT)**
When does the partial eclipse end on the centerline at the Kansas/Missouri border? **2:34 p.m. (CDT)**

When does the partial eclipse start on the centerline at the Missouri/Illinois border? **11:51 a.m. (CDT)**
When does totality start on the centerline at the Missouri/Illinois border? **1:18 p.m. (CDT)**
When does the partial eclipse end on the centerline at the Missouri/Illinois border? **2:46 p.m. (CDT)**

Check local forecasts and satellite maps for the best weather information for this eclipse.

Cities and Highways in Totality by Location

1-25: Northwest Missouri - **I-29** - Craig, Mound City, Oregon, Amazonia, Savannah, Helena, St. Joseph

26-70: Northwest Missouri - **I-29, US-169, US-69, I-35, MO-36, MO-33, MO-13, MO-210, MO-10** Weston, Platte City, Kansas City (far north), Smithville, Liberty, Excelsior Springs, Gower, Edgerton, Plattsburg, Lathrop, Kearney, Lawson, Sibley, Richmond, Lexington

71-95: Central Missouri - **MO-13, US-65, I-70, MO-41, MO-5, MO-40** - Higginsville, Concordia, Sweet Springs, Carrollton, Waverly, Marshall, Arrow Rock, Glasgow, Boonville

96-131: Central Missouri - **I-70, US-50, US-63, US-54, MO-94** - Tipton, Rocheport, Columbia, Fulton, Ashland, California, Jefferson City, Westphalia

132-160: Eastern Missouri - **US-50, MO-94, MO-47, I-44, MO-30** - Hermann, New Haven, Sullivan, Union, Washington, Robertsville, Eureka, Ballwin, Cedar Hill

161-190: Eastern Missouri - **MO-21, I-55, US-61, US-67** - Arnold, Imperial, Hillsboro, Festus, De Soto, Bonne Terre, Farmington, Ste. Genevieve, Perryville, Jackson

Weather and Eclipse Related Links

Website

NW Missouri Weather
Pleasant Hill, MO NWS Office
http://www.weather.gov/eax/

St. Louis 2017 Eclipse
http://www.stlouiseclipse2017.org

Eastern Missouri Weather
St. Louis, MO NWS Office
http://www.weather.gov/lsx/

St. Joseph 2017 Eclipse
http://stjosepheclipse.com/start.html

Midwest U.S. GOES Satellite - Visible Loop
http://www.ssd.noaa.gov/goes/east/mw/h5-mloop-vis.html

Big Lake State Park adjoins the largest oxbow lake in Missouri and offers boating, swimming, camping and picnic areas. The total eclipse lasts two minutes and thirty one seconds at this state park and starts at 1:05 p.m. CDT. See Missouri #4 for location details.

Website

Photos

Arrow Rock State Historic Site lets you step back in time on eclipse day at the oldest continuously operating restaurant west of the Mississippi. Visitors will experience two minutes and forty seconds of totality, basically the maximum duration for this eclipse. Campgrounds are available. Totality starts at 1:10 p.m. CDT. See Missouri #90 for details.

Website

Photos

Weston Bend State Park boasts a commanding view of the western horizon overlooking the Missouri River. The scenic overlook, picnic sites and hiking trails make this a tempting destination for eclipse watchers. Totality starts at 1:07 p.m. CDT. See Missouri #27 for details.

Website

Photos

Robertsville State Park offers many outdoor activities including camping, fishing and boating along the Meramec River. Over two and a half minutes of totality will darken the skies at this state park on eclipse day. Totality starts at 1:16 p.m. CDT. See Missouri #154 for location details.

Website

Photos

MO Crossing Points Table

Where the Eclipse Centerline Crosses Highways in Missouri

Start Times = Central Daylight Time

Loc#	Hwy	Nearest Mile Marker, Cross Street or Exit, City	TStart	TLasts
24	I-29	Exit 44, US-169, St. Joseph, MO	1:06	2:39
53	MO-33	NW Bretheren Church Rd, Plattsburg, MO	1:07	2:39
56	I-35	MM 39.4, Exit 40, MO-116, Lathrop, MO	1:07	2:39
57	US-69	SE 240th St, Lathrop, MO	1:07	2:39
67	MO-13	E 206th St, Rayville, MO	1:08	2:39
78	MO-10	Co Rd 617/519, Norborne, MO	1:09	2:40
89	MO-41	Pleasant Ave, Marshall, MO	1:10	2:40
94	MO-5	Jct MO-5/US-40, New Franklin, MO	1:11	2:40
100	I-70	MM 111.8, Exit 111, Boonville, MO	1:11	2:40
116	US-63	Minor Hill Rd / Angel Ln, Columbia, MO	1:12	2:40
122	US-54	2.7 Miles south of State Rd BB, New Bloomfield, MO	1:13	2:40
131	MO-94	Fulton Ave, Mokane, MO	1:13	2:40
139	MO-19	1.2 miles south of Hwy E, Boeuf Township, MO	1:14	2:41
144	US-50	Flint Hill Rd, Leslie, MO	1:15	2:41
152	US-66/I-44	I-44 MM 238.2, Weigh Station, St. Clair, MO	1:15	2:41
153	MO-47	Kommer Loop, Prairie, MO	1:16	2:41
166	MO-21	City limit sign, .5 miles south of MO-110, De Soto, MO	1:16	2:41
171	US-67	.5 miles south of MO-110, De Soto, MO	1:17	2:41
178	I-55	MM 161.6, Exit 162, Bloomsdale, MO	1:17	2:41
180	US-61	Industrial River Rd, Ste. Genevieve, MO	1:17	2:41
183	MO-51	.7 miles south of PCR-238, Perryville, MO	1:18	2:41

Northwest Missouri

Scan for Map   **Start Times = Central Daylight Time**

9 Davis Roadside Park
Savannah, MO 64485
39.96589 -94.95265
Totality Starts at 1:06 / Lasts 2:27

1 Thurnau Conservation Area
Austin Rd, Craig, MO 64437
40.16920 -95.45731
Totality Starts at 1:04 / Lasts 2:23

10 Savannah City Park
1398 W Chestnut St
Savannah, MO 64485
39.94337 -94.84132
Totality Starts at 1:06 / Lasts 2:26

2 Sports Complex
Nebraska St, Mound City, MO 64470
40.14887 -95.23213
Totality Starts at 1:05 / Lasts 2:15

11 Andrew County Courthouse
411 Court St Savannah, MO 64485
39.94124 -94.82989
Totality Starts at 1:06 / Lasts 2:26

3 Chautauqua Park
499 E Gillis St, Mound City, MO 64470
40.13668 -95.22958
Totality Starts at 1:05 / Lasts 2:17

12 Savannah Sports Complex
N Ironwood, Savannah, MO 64485
39.94111 -94.80493
Totality Starts at 1:06 / Lasts 2:24

4 Big Lake State Park
204 Lake Shore Dr, Craig, MO 64437
40.05314 -95.36317
Totality Starts at 1:05 / Lasts 2:31

13 Helena City Park
3rd St Helena, MO 64459
39.91311 -94.65210
Totality Starts at 1:06 / Lasts 2:21

5 Brown Conservation Area
Forest City, MO 64451
39.96005 -95.24734
Totality Starts at 1:05 / Lasts 2:34

14 Huston Wyeth Park
St Joseph, MO 64501
39.77691 -94.86677
Totality Starts at 1:06 / Lasts 2:38

6 Welty Park
Oregon, MO 64473
39.98475 -95.13700
Totality Starts at 1:05 / Lasts 2:31

15 Krug Park
St Joseph, MO 64505
39.80205 -94.85128
Totality Starts at 1:06 / Lasts 2:36

7 Riverbreaks Conservation Area
Oregon, MO 64473
39.93967 -95.14117
Totality Starts at 1:05 / Lasts 2:34

16 Heritage Park Softball Complex
2202 Waterworks Rd
St Joseph, MO 64501
39.78905 -94.87520
Totality Starts at 1:06 / Lasts 2:37

8 Honey Creek Conservation Area
Amazonia, MO 64421
39.95276 -94.97032
Totality Starts at 1:06 / Lasts 2:29

17 St Joseph Nature Center
1502 McArthur Dr, St Joseph, MO 64505
39.77975 -94.87393
Totality Starts at 1:06 / Lasts 2:38

Plan to be at your eclipse viewing site at least two hours before totality starts.

18 Roserans Memorial Airport
A Ave, St Joseph, MO 64503
39.77282 -94.90152
Totality Starts at 1:06 / Lasts 2:38

19 France Park
N 26th St, St Joseph, MO 64501
39.77166 -94.82751
Totality Starts at 1:06 / Lasts 2:37

20 Youngdahl Urban CA
N 32nd St, St Joseph, MO 64501
39.76593 -94.81203
Totality Starts at 1:06 / Lasts 2:37

21 Lewis Park
Vine St, St Joseph, MO 64501
39.76070 -94.84572
Totality Starts at 1:06 / Lasts 2:38

22 Patee Park
Penn St, St Joseph, MO 64503
39.75687 -94.84867
Totality Starts at 1:06 / Lasts 2:38

23 Bartlett Park
Monterey St, St Joseph, MO 64503
39.75377 -94.81894
Totality Starts at 1:06 / Lasts 2:38

24 CP I-29
Exit 44, St Joseph, MO 64503
39.72110 -94.78841
Totality Starts at 1:06 / Lasts 2:39

25 Hyde Park
402 East Hyde Park Avenue
St. Joseph, MO 64504
39.71055 -94.85009
Totality Starts at 1:06 / Lasts 2:39

26 Lewis and Clark State Park
Lakecrest Blvd, Rushville, MO 64484
39.53818 -95.05749
Totality Starts at 1:06 / Lasts 2:18

27 Weston Bend State Park
16600 MO-45, Weston, MO 64098
39.39053 -94.87798
Totality Starts at 1:07 / Lasts 2:02

28 Platte Ridge Park
17130 MO-371, Platte City, MO 64079
39.40251 -94.80126
Totality Starts at 1:07 / Lasts 2:09

29 Platte County Fairgrounds
Platte City, MO 64079
39.38096 -94.79021
Totality Starts at 1:07 / Lasts 2:06

30 Riverview Park
298 W Mill St, Platte City, MO 64079
39.36185 -94.79449
Totality Starts at 1:07 / Lasts 2:02

31 Belcher Branch Lake CA
Faucett, MO 64448
39.58424 -94.73430
Totality Starts at 1:06 / Lasts 2:37

32 Gower City Park
S 3rd St Gower, MO 64454
39.61187 -94.60006
Totality Starts at 1:07 / Lasts 2:39

33 Veterans Memorial
N Clark Ave, Edgerton, MO 64444
39.50610 -94.62897
Totality Starts at 1:07 / Lasts 2:34

34 Heritage Park
320 E Main St Smithville, MO 64089
39.38835 -94.57658
Totality Starts at 1:07 / Lasts 2:20

35 Smith Fork Park Campground
601 Co Rd DD, Smithville, MO 64089
39.39634 -94.55915
Totality Starts at 1:07 / Lasts 2:23

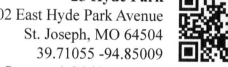

Plan to be at your eclipse viewing site at least two hours before totality starts.

Scan for Map

36 Crows Creek Campground
Smithville Lake, Smithville, MO 64089
39.39500 -94.52786
Totality Starts at 1:07 / Lasts 2:24

37 Clay County Park
17600 Collins Rd Smithville, MO
39.41145 -94.52159
Totality Starts at 1:07 / Lasts 2:27

38 Anne Garney Park
N Woodland Ave, Kansas City, MO
39.29532 -94.55873
Totality Starts at 1:07 / Lasts 2:05

39 Hodge Park
7000 N.E. Barry Rd, Kansas City, MO
39.26034 -94.49894
Totality Starts at 1:07 / Lasts 2:03

40 Stocksdale Park
901 S La Frenz Rd, Liberty, MO 64068
39.23380 -94.38949
Totality Starts at 1:08 / Lasts 2:05

41 Fountain Bluff Sports Complex
2200 Old State Hwy 210, Liberty, MO
39.20492 -94.38427
Totality Starts at 1:08 / Lasts 2:00

42 Fort Osage Historic Landmark
107 Osage St, Sibley, MO 64088
39.18826 -94.19369
Totality Starts at 1:08 / Lasts 2:09

43 Hayes Park
Santa Fe St, Sibley, MO 64088
39.18264 -94.19555
Totality Starts at 1:08 / Lasts 2:08

44 Stewartsville City Park
8th St, Stewartsville, MO 64490
39.75277 -94.49734
Totality Starts at 1:07 / Lasts 2:32

Scan for Map

45 Pony Express Lake CA
Osborn, MO 64474
39.80109 -94.37973
Totality Starts at 1:07 / Lasts 2:24

46 Osborn City Park
Hunt St, Osborn, MO 64474
39.74994 -94.35861
Totality Starts at 1:07 / Lasts 2:29

47 Cameron Aquatic Center
Seminary Ave, Cameron, Missouri
39.73309 -94.24220
Totality Starts at 1:07 / Lasts 2:27

48 Cameron City Park
N Cherry St, Cameron, MO 64429
39.74203 -94.23907
Totality Starts at 1:07 / Lasts 2:26

49 Hamilton City Park
S Hughes St, Hamilton, MO 64644
39.74098 -94.00118
Totality Starts at 1:08 / Lasts 2:13

50 Wallace State Park
10621 MO-121 Cameron, MO 64429
39.65791 -94.21467
Totality Starts at 1:07 / Lasts 2:32

51 Smithville Lake St. Wildlife Area
Plattsburg, MO 64477
39.52970 -94.48710
Totality Starts at 1:07 / Lasts 2:38

52 Perkins Park
Plattsburg, MO 64477
39.56306 -94.44185
Totality Starts at 1:07 / Lasts 2:39

53 CP MO-33
Plattsburg, MO 64477
39.59228 -94.41283
Totality Starts at 1:07 / Lasts 2:39

Plan to be at your eclipse viewing site at least two hours before totality starts.

Scan for Map

54 Hartell Conservation Area
Turney, MO 64493
39.60315 -94.39739
Totality Starts at 1:07 / Lasts 2:39

55 Lathrop City Park
Park St, Lathrop, MO 64465
39.54694 -94.33327
Totality Starts at 1:07 / Lasts 2:40

56 CP I-35
Exit 40, Lathrop, MO 64465
39.54382 -94.27290
Totality Starts at 1:07 / Lasts 2:39

57 CP 275 US-69
Lathrop, MO 64465
39.53104 -94.23613
Totality Starts at 1:07 / Lasts 2:39

58 Lake Arrowhead
Mohawk Dr Lathrop, MO 64465
39.48492 -94.32353
Totality Starts at 1:07 / Lasts 2:39

59 Jesse James Park
3001 N Hwy 33 Kearney, MO 64060
39.39847 -94.36550
Totality Starts at 1:07 / Lasts 2:32

60 Lions Park
340 S Jefferson St, Kearney, MO 64060
39.37159 -94.36318
Totality Starts at 1:07 / Lasts 2:29

61 Lawson City Park
W 4th St, Lawson, MO 64062
39.43792 -94.20486
Totality Starts at 1:08 / Lasts 2:39

62 Watkins Mill State Park
26600 Park Rd N Lawson, MO 64062
39.40105 -94.25443
Totality Starts at 1:08 / Lasts 2:36

Scan for Map

63 Century Park
112 S Thompson Ave
Excelsior Springs, MO 64024
39.34542 -94.24546
Totality Starts at 1:08 / Lasts 2:31

64 Paul Craig Park
Isley Blvd, Excelsior Springs, MO
39.34153 -94.22062
Totality Starts at 1:08 / Lasts 2:32

65 Siloam Mountain Park
Excelsior Springs, MO 64024
39.33846 -94.22118
Totality Starts at 1:08 / Lasts 2:31

66 Tait Memorial Park
SE Tait Park Dr, Braymer, MO 64624
39.58279 -93.78432
Totality Starts at 1:08 / Lasts 2:26

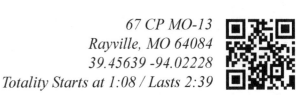

67 CP MO-13
Rayville, MO 64084
39.45639 -94.02228
Totality Starts at 1:08 / Lasts 2:39

68 Ray County Lake
Richmond, MO 64085
39.31336 -93.99523
Totality Starts at 1:08 / Lasts 2:37

69 Roberts Park
Richmond, MO 64085
39.27742 -93.96259
Totality Starts at 1:08 / Lasts 2:35

70 Battle of Lexington Historic Site
1101 Delaware St Lexington, MO
39.19352 -93.87908
Totality Starts at 1:09 / Lasts 2:29

71 Confederate Memorial Site
211 W 1st St Higginsville, MO 64037
39.09659 -93.72411
Totality Starts at 1:09 / Lasts 2:22

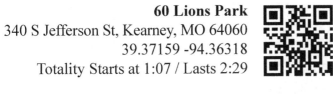

Plan to be at your eclipse viewing site at least two hours before totality starts.

72 Fairground Park
Higginsville, MO 64037
39.06695 -93.73061
Totality Starts at 1:09 / Lasts 2:17

81 Grand Pass Conservation Area
Highway 65 Miami, MO 65344
39.28980 -93.31367
Totality Starts at 1:10 / Lasts 2:35

73 Concordia Rest Area
I-70, Concordia, MO 64020
38.99401 -93.58581
Totality Starts at 1:09 / Lasts 2:13

82 Van Meter State Park
Miami, MO 65344
39.27558 -93.26028
Totality Starts at 1:10 / Lasts 2:35

74 Central Park
Concordia, MO 64020
38.98280 -93.56892
Totality Starts at 1:09 / Lasts 2:12

83 Blind Pony Lake CA
Sweet Springs, MO 65351
39.04500 -93.36137
Totality Starts at 1:10 / Lasts 2:33

75 South Side Park
S Main St, Concordia, MO 64020
38.97417 -93.57026
Totality Starts at 1:10 / Lasts 2:10

Central Missouri

76 Sweet Springs City Park
Columbia Ave, Sweet Springs, MO
38.94909 -93.41928
Totality Starts at 1:10 / Lasts 2:16

84 Marshall Municipal Airport
Marshall, MO 65340
39.10184 -93.20028
Totality Starts at 1:10 / Lasts 2:39

77 Carrollton Recreation Park
603 Chillicothe Rd, Carrollton, MO
39.36623 -93.48867
Totality Starts at 1:09 / Lasts 2:35

85 Marshall Junction RV Park
Marshall, MO 65340
38.96215 -93.19887
Totality Starts at 1:10 / Lasts 2:31

78 CP MO-10
Norborne, MO 64668
39.31506 -93.62201
Totality Starts at 1:09 / Lasts 2:40

86 Indian Foothills Park
Marshall, MO 65340
39.12059 -93.16721
Totality Starts at 1:10 / Lasts 2:40

79 CP US-24
Carrollton, MO 64633
39.27593 -93.51225
Totality Starts at 1:09 / Lasts 2:40

87 CP MO-240
Marshall, MO 65340
39.12915 -93.10444
Totality Starts at 1:10 / Lasts 2:40

80 Missouri River Park
W Thomas Dr, Waverly, MO 64096
39.21395 -93.51048
Totality Starts at 1:09 / Lasts 2:39

88 Slater Lake
30399 Nodaway Dr Slater, MO 65349
39.22007 -93.04944
Totality Starts at 1:10 / Lasts 2:34

Plan to be at your eclipse viewing site at least two hours before totality starts.

89 CP MO-41
Marshall, MO 65340
39.08938 -92.99492
Totality Starts at 1:10 / Lasts 2:40

98 Moniteau Co Fairgrounds
700 E South St, California, MO 65018
38.63405 -92.56112
Totality Starts at 1:12 / Lasts 2:16

90 Arrow Rock State Historic Site
Nelson, MO 65347
39.06433 -92.94264
Totality Starts at 1:10 / Lasts 2:40

99 Prairie Home Conservation Area
Prairie Rd, Prairie Home, MO 65068
38.77994 -92.60057
Totality Starts at 1:12 / Lasts 2:35

91 Stump Island Recreation Park
Glasgow, MO 65254
39.21874 -92.84815
Totality Starts at 1:11 / Lasts 2:29

100 CP I-70
Exit 111, Boonville, MO 65233
38.93995 -92.58730
Totality Starts at 1:11 / Lasts 2:40

92 Fayette City Park
Fayette, MO 65248
39.15140 -92.69406
Totality Starts at 1:11 / Lasts 2:31

101 Diana Bend Conservation Area
New Franklin, MO 65274
38.97781 -92.57329
Totality Starts at 1:11 / Lasts 2:38

93 Katy Roundhouse Campground
New Franklin, MO 65274
39.01167 -92.74900
Totality Starts at 1:11 / Lasts 2:40

102 Tribble Park
State Hwy OO, Hallsville, MO 65255
39.11564 -92.22013
Totality Starts at 1:12 / Lasts 2:12

94 CP MO-5
New Franklin, MO 65274
38.99736 -92.74313
Totality Starts at 1:11 / Lasts 2:40

103 Finger Lakes State Park
Peabody Rd, Columbia, MO 65202
39.08768 -92.31636
Totality Starts at 1:12 / Lasts 2:22

95 Rolling Hills Park
Boonville, MO 65233
38.96082 -92.75373
Totality Starts at 1:12 / Lasts 2:39

104 Atkins Memorial Park
Columbia, MO 65202
39.00993 -92.29536
Totality Starts at 1:12 / Lasts 2:30

96 Tipton City Park
S Ferguson Ave, Tipton, MO 65081
38.64548 -92.77788
Totality Starts at 1:12 / Lasts 2:04

105 Cottonwood RV Park
5170 N Oakland Gravel Rd
Columbia, MO 65202
39.00687 -92.30091
Totality Starts at 1:12 / Lasts 2:31

97 Proctor Park Lake
Parkway Dr, California, MO 65018
38.62133 -92.56372
Totality Starts at 1:12 / Lasts 2:13

106 Cosmo Park
Business Loop 70 W, Columbia, MO
38.97067 -92.36314
Totality Starts at 1:12 / Lasts 2:34

Plan to be at your eclipse viewing site at least two hours before totality starts.

107 Columbia Mall
Bernadette Dr, Columbia, MO 65203
38.96237 -92.37682
Totality Starts at 1:12 / Lasts 2:35

108 Stephens Lake Park
E Broadway, Columbia, MO 65201
38.95157 -92.30654
Totality Starts at 1:12 / Lasts 2:34

109 Longview Park
4980 W Gillespie Bridge Rd
Columbia, MO 65203
38.92949 -92.41014
Totality Starts at 1:12 / Lasts 2:38

110 Twin Lakes Recreation Area
Columbia, MO 65203
38.92754 -92.37863
Totality Starts at 1:12 / Lasts 2:37

111 Grindstone Nature Area
2435 Old 63 S, Columbia, MO 65201
38.92620 -92.31177
Totality Starts at 1:12 / Lasts 2:36

112 Cosmo-Bethel Park
4500 Bethel St, Columbia, MO 65203
38.90253 -92.34258
Totality Starts at 1:12 / Lasts 2:38

113 Rock Bridge State Park
5901 South Highway 163
Columbia, MO 65203
38.87326 -92.32576
Totality Starts at 1:12 / Lasts 2:39

114 Katfish Katy Campgrounds
8825 Sarr St, Columbia, MO 65203
38.91067 -92.47376
Totality Starts at 1:12 / Lasts 2:40

115 Eagle Bluffs Overlook
Katy Trail, Columbia, MO 65203
38.85031 -92.41964
Totality Starts at 1:12 / Lasts 2:40

116 CP 11625 US-63
Columbia, MO 65201
38.81512 -92.25128
Totality Starts at 1:12 / Lasts 2:40

117 Columbia Regional Airport
S Airport Dr, Columbia, MO 65201
38.80550 -92.22359
Totality Starts at 1:12 / Lasts 2:40

118 Ashland City Park
399 N College St, Ashland, MO 65010
38.77744 -92.25882
Totality Starts at 1:12 / Lasts 2:40

119 Newman Lake
3299 Pinetree Dr, Columbia, MO 65201
38.94529 -92.14476
Totality Starts at 1:12 / Lasts 2:31

120 Little Dixie Lake CA
Fulton, MO 65251
38.91819 -92.12045
Totality Starts at 1:12 / Lasts 2:32

121 Memorial Park
Fulton, MO 65251
38.84416 -91.94941
Totality Starts at 1:13 / Lasts 2:33

122 CP US-54
New Bloomfield, MO 65063
38.74220 -92.05682
Totality Starts at 1:13 / Lasts 2:40

123 Binder State Park
Binder Lake Rd, Jefferson City, MO
38.60175 -92.30585
Totality Starts at 1:12 / Lasts 2:27

124 Noren River Access
Jefferson City, MO 65101
38.58997 -92.17761
Totality Starts at 1:13 / Lasts 2:31

Plan to be at your eclipse viewing site at least two hours before totality starts.

Scan for Map Scan for Map

125 Jefferson City Memorial Airport
Airport Rd, Jefferson City, MO 65101
38.59202 -92.15615
Totality Starts at 1:13 / Lasts 2:33

134 Steamboat Junction Campground
199 MO-94 Rhineland, MO 65069
38.70614 -91.62247
Totality Starts at 1:14 / Lasts 2:34

126 Washington Park
Jefferson City, MO 65109
38.57632 -92.18882
Totality Starts at 1:13 / Lasts 2:29

135 Grand Bluffs CA
Rhineland, MO 65069
38.70630 -91.60863
Totality Starts at 1:14 / Lasts 2:33

127 Ellis-Porter Park
Jefferson City, MO 65101
38.56754 -92.14445
Totality Starts at 1:13 / Lasts 2:30

136 Gasconade Park
Gasconade, MO 65061
38.66679 -91.55871
Totality Starts at 1:14 / Lasts 2:34

128 Hough Park Lake
Jefferson City, MO 65101
38.53995 -92.18399
Totality Starts at 1:13 / Lasts 2:24

137 Hermann City Park
Gasconade St, Hermann, MO 65041
38.69833 -91.44135
Totality Starts at 1:14 / Lasts 2:29

129 Painted Rock Conservation Area
MO-133, Westphalia, MO 65085
38.39838 -92.11236
Totality Starts at 1:13 / Lasts 2:04

138 Riverfront Park
Wharf St, Hermann, MO 65041
38.70783 -91.43292
Totality Starts at 1:14 / Lasts 2:28

130 Clark's Hill Historic Site
Jefferson City, MO 65101
38.55844 -92.02984
Totality Starts at 1:13 / Lasts 2:34

139 CP MO-19
Boeuf Township, MO 65041
38.51566 -91.46087
Totality Starts at 1:14 / Lasts 2:41

131 CP MO-94
Mokane, MO 65059
38.67215 -91.87120
Totality Starts at 1:13 / Lasts 2:40

140 Lions Park
Owensville, MO 65066
38.35402 -91.48909
Totality Starts at 1:14 / Lasts 2:35

132 Ben Branch Lake CA
Chamois, MO 65024
38.57018 -91.78918
Totality Starts at 1:13 / Lasts 2:40

141 Lost Valley Lake Resort
2334 Hwy ZZ, Owensville, MO 65066
38.47499 -91.38921
Totality Starts at 1:14 / Lasts 2:41

133 Graham Cave State Park
Montgomery City, MO 63361
38.90776 -91.57705
Totality Starts at 1:14 / Lasts 2:05
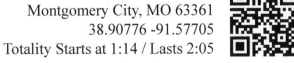

142 Gerald City Park
Gerald, MO 63037
38.39181 -91.31960
Totality Starts at 1:15 / Lasts 2:40

Plan to be at your eclipse viewing site at least two hours before totality starts.

Southeast Missouri

143 Lions Field Park
101 Lions Dr, New Haven, MO 63068
38.59876 -91.22919
Totality Starts at 1:15 / Lasts 2:31

144 CP 4364 US-50
Leslie, MO 63056
38.42147 -91.21669
Totality Starts at 1:15 / Lasts 2:41

145 City Lake Park
Mattox Dr, Sullivan, MO 63080
38.19349 -91.17671
Totality Starts at 1:15 / Lasts 2:30

146 Meramec State Park
670 Fisher Cave Dr, Sullivan, MO
38.20768 -91.09652
Totality Starts at 1:15 / Lasts 2:35

147 Stanton / Meramec KOA
74 Hwy W Sullivan, MO 63080
38.27338 -91.10236
Totality Starts at 1:15 / Lasts 2:39

148 Washington City Park
High St, Washington, MO 63090
38.56417 -91.02578
Totality Starts at 1:15 / Lasts 2:27

149 Burger Park
Int'l Ave, Washington, MO 63090
38.54024 -90.98615
Totality Starts at 1:15 / Lasts 2:28

150 Union City Park
Union, MO 63084
38.44836 -91.01809
Totality Starts at 1:15 / Lasts 2:35

151 Orchard Park
Orchard Dr, St Clair, MO 63077
38.36432 -90.97242
Totality Starts at 1:15 / Lasts 2:39

152 CP Historic US-66
St Clair, MO 63077
38.34008 -91.00725
Totality Starts at 1:15 / Lasts 2:41

153 CP 2220 MO-47
Prairie, MO 63077
38.28640 -90.86989
Totality Starts at 1:16 / Lasts 2:41

154 Robertsville State Park
State Park Rd, Robertsville, MO 63072
38.42261 -90.82069
Totality Starts at 1:16 / Lasts 2:32

155 Six Flags St. Louis
4900 Six Flags Rd, Eureka, MO 63025
38.50851 -90.67700
Totality Starts at 1:16 / Lasts 2:15

156 Lions Park
340 Bald Hill Rd, Eureka, MO 63025
38.49926 -90.62961
Totality Starts at 1:16 / Lasts 2:14

157 Route 66 State Park
97 N Outer Rd E #1, Eureka, MO 63025
38.50270 -90.59249
Totality Starts at 1:16 / Lasts 2:10

158 Sherman Beach Park
St Paul Rd, Ballwin, MO 63021
38.53589 -90.58795
Totality Starts at 1:16 / Lasts 2:02

159 Byrnes Mill Park
Osage Exec Cir, Byrnes Mill, MO 63051
38.43790 -90.58314
Totality Starts at 1:16 / Lasts 2:20

Plan to be at your eclipse viewing site at least two hours before totality starts.

Scan for Map

160 Cedar Hill Park
Cedar Hill Rd Cedar Hill, MO 63016
38.34917 -90.64434
Totality Starts at 1:16 / Lasts 2:33

161 Sandy Creek Covered Bridge State Historic Site
Old Lemay Ferry Rd, Hillsboro, MO
38.29450 -90.52560
Totality Starts at 1:16 / Lasts 2:33

162 Arnold City Park
Bradley Beach Rd, Arnold, MO 63010
38.45290 -90.35321
Totality Starts at 1:17 / Lasts 2:00

163 Mastodon State Historic Site
Charles J Becker Drive, Imperial, MO
38.37937 -90.39344
Totality Starts at 1:17 / Lasts 2:18

164 Teamster Camp and Picnic Gardens
Teamster Lake, Pevely, MO 63070
38.29016 -90.38010
Totality Starts at 1:17 / Lasts 2:29

165 Sunset Park
Festus, MO 63028
38.22139 -90.39984
Totality Starts at 1:17 / Lasts 2:34

166 CP 12510 MO-21
De Soto, MO 63020
38.16715 -90.56704
Totality Starts at 1:16 / Lasts 2:41

167 Ritcher Park
De Soto, MO 63020
38.13797 -90.55052
Totality Starts at 1:16 / Lasts 2:41

168 Spross Park
De Soto, MO 63020
38.13490 -90.56388
Totality Starts at 1:16 / Lasts 2:41

Scan for Map

169 Hopson Field
Lake St De Soto, MO 63020
38.12812 -90.55672
Totality Starts at 1:16 / Lasts 2:41

170 Washington State Park
13041 MO-104 De Soto, MO 63020
38.08896 -90.69742
Totality Starts at 1:16 / Lasts 2:39

171 CP 4051 U.S. 67
De Soto, MO 63020
38.12946 -90.47198
Totality Starts at 1:17 / Lasts 2:41

172 St. Francois State Park
8920 US-67 Bonne Terre, MO 63628
37.96868 -90.53374
Totality Starts at 1:17 / Lasts 2:34

173 Bonne Terre City Park
S Park St, Bonne Terre, MO 63628
37.92752 -90.55999
Totality Starts at 1:17 / Lasts 2:27

174 Desloge City Park
W Walnut St, Desloge, MO 63601
37.86834 -90.53097
Totality Starts at 1:17 / Lasts 2:19

175 Missouri Mines Historic Site
4000 MO-32, Park Hills, MO 63601
37.83750 -90.51091
Totality Starts at 1:17 / Lasts 2:15

176 Wilson Rozier Park
Perrine Rd, Farmington, MO 63640
37.77532 -90.43136
Totality Starts at 1:17 / Lasts 2:10

177 Hawn State Park
12096 Park Dr, Ste. Genevieve, MO
37.83371 -90.24234
Totality Starts at 1:17 / Lasts 2:32

Plan to be at your eclipse viewing site at least two hours before totality starts.

Scan for Map

178 CP I-55
Bloomsdale, MO 63627
38.04880 -90.26949
Totality Starts at 1:17 / Lasts 2:41

179 Magnolia Hollow CA
Ste. Genevieve, MO 63670
38.03363 -90.12409
Totality Starts at 1:17 / Lasts 2:39

180 CP US-61
Ste. Genevieve, MO 63670
37.97896 -90.09522
Totality Starts at 1:17 / Lasts 2:41

181 Pere Marquette Park
Ste. Genevieve, MO 63670
37.99498 -90.05562
Totality Starts at 1:18 / Lasts 2:39

182 CP 19344 US-61
Ste. Genevieve, MO 63670
37.94694 -90.01563
Totality Starts at 1:18 / Lasts 2:41

183 CP 11477 MO-51
Perryville, MO 63775
37.87282 -89.83223
Totality Starts at 1:18 / Lasts 2:41

184 Perry County Community Lake
Perryville, MO 63775
37.72258 -89.90966
Totality Starts at 1:18 / Lasts 2:34

185 Perry Park Center
City Park Ln, Perryville, MO 63775
37.72507 -89.85272
Totality Starts at 1:18 / Lasts 2:37

186 Seventy-Six Conservation Area
Frohna, MO 63748
37.72168 -89.61964
Totality Starts at 1:19 / Lasts 2:40

Scan for Map

187 Tower Rock Natural Area
Pcr 460, Frohna, MO 63748
37.63106 -89.51646
Totality Starts at 1:19 / Lasts 2:39

188 Apple Creek Conservation Area
Jackson, MO 63755
37.55076 -89.56616
Totality Starts at 1:19 / Lasts 2:31

189 Weiss Roadside Park
I-55, Jackson, MO 63755
37.42984 -89.64208
Totality Starts at 1:19 / Lasts 2:06

190 Trail of Tears State Park
Moccasin Springs Rd
Jackson, MO 63755
37.45448 -89.46346
Totality Starts at 1:20 / Lasts 2:23

Plan to be at your eclipse viewing site at least two hours before totality starts.

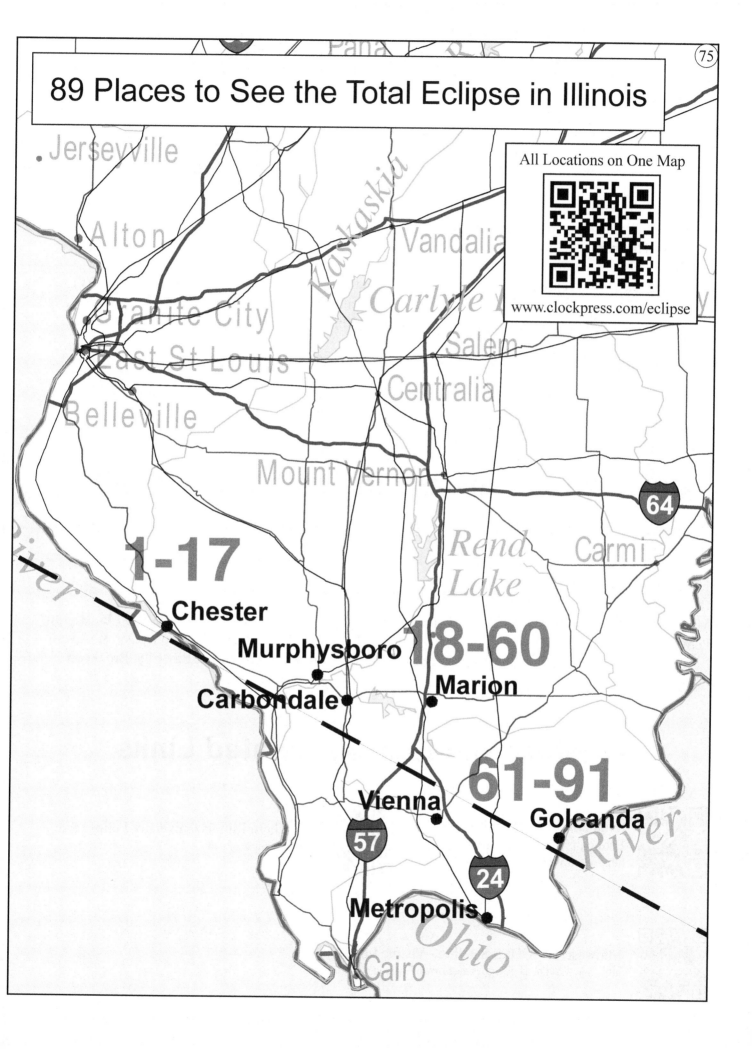

Illinois Eclipse Track Notes

How many miles long is the eclipse centerline in Illinois? **76 miles**

What is the average duration of totality along the centerline in Illinois? **2:41 (m:ss)**
How long does totality last on the centerline at the Missouri/Illinois border? **2:41 (m:ss)**
How long does totality last on the centerline at the Illinois/Kentucky border? **2:41 (m:ss)**

When does the partial eclipse start on the centerline at the Missouri/Illinois border? **11:51 a.m. (CDT)**
When does totality start on the centerline at the Missouri/Illinois border? **1:18 p.m. (CDT)**
When does the partial eclipse end on the centerline at the Missouri/Illinois border? **2:46 p.m. (CDT)**

When does the partial eclipse start on the centerline at the Illinois/Kentucky border? **11:54 a.m. (CDT)**
When does totality start on the centerline at the Illinois/Kentucky border? **1:22 p.m. (CDT)**
When does the partial eclipse end on the centerline at the Illinois/Kentucky border? **2:49 p.m. (CDT)**

Check local forecasts and satellite maps for the best weather information for this eclipse.

Cities and Highways in Totality by Location

1-14: South of I-64 - **IL-3, IL-159, IL-154, IL-4** Waterloo, Redbud, Sparta, Grove, Chester

15-50: West of I-57 - **IL-3, IL-4, IL-13, US-51, IL-149, IL-127** - Pinckneyville, DuQuoin, Murphysboro, Carbondale, Makanda, Jonesboro, Pomona

51-70: I-57 & I-24 Corridor - **I-57, I-24, IL-149, IL-148, IL-13, IL-37, IL-146, US-45** - Royalton, Herrin, Johnston City, Marion, Ozark, Goreville, Vienna

71-89: I-24 and East - **I-24, US-45, IL-145, IL-146, IL-147, IL-1** - Mermet, Metropolis, Derby, Cave-in-Rock, Rosiclare, Golconda, Bay City, Brookport

Weather and Eclipse Related Links

Website

Southern Illinois Weather
Paducah, KY NWS Office
http://www.weather.gov/pah/

Southern Illinois University Eclipse Site
http://eclipse.siu.edu

Midwest U.S. GOES Satellite - Visible Loop
http://www.ssd.noaa.gov/goes/east/mw/h5-mloop-vis.html

Featured Eclipse Destinations

Fort Kaskaskia was established over 300 years ago. A large campground, picnic area, historic cemetery and scenic overlook of the Mississippi River make this a great place to see the eclipse. Almost two minutes and forty seconds of totality will be seen from this site. Totality starts at 1:18 p.m. CDT. See Illinois #8 for location details.

Website

Photos

Dixon Springs State Park has camping and picnic areas, hiking trails and much more on 800 acres. Being very close to the centerline gives this site the full two minutes and forty one seconds of totality on eclipse day. Totality starts at 1:21 p.m. CDT. See Illinois #78 for details.

Website

Photos

Lake Murphysboro State Park is over 1,000 acres of outdoor activities including camping, archery, fishing and more. The total eclipse of the sun will last over two and a half minutes at this scenic location. Totality starts at 1:19 p.m. CDT. See Illinois #23 for details.

Website

Photos

Fort Massac State Park on the Ohio River in southern Illinois shows off a historic replica fort on 1,500 acres. Boating and swimming, camping and picnic areas are a few of the attractions here. Nearly two and a half minutes of totality will be seen from here on eclipse day. Totality starts at 1:21 p.m. CDT. See Illinois #81 for location details.

Website

Photos

IL Crossing Points Table

Where the Eclipse Centerline Crosses Highways in Illinois

Start Times = Central Daylight Time

Loc#	Hwy	Nearest Mile Marker, Cross Street or Exit, City	TStart	TLasts
25	IL-3	Town Creek Rd, Gorham, IL	1:19	2:41
35	IL-127	Tomcat Hill Ln, Carbondale, IL	1:19	2:41
42	US-51	Curry Ln, Makanda, IL	1:20	2:41
61	I-57	MM 39, Exit 40, Buncombe, IL	1:20	2:41
63	IL-37	Tall Tree Lake Rd, Goreville, IL	1:21	2:41
65	I-24	MM 12, Vienna, IL	1:21	2:41
66	US-45	Roosevelt Loop, Vienna, IL	1:21	2:41
77	IL-145	Franks Road, Golconda, IL	1:21	2:41
79	IL-146	2.2 miles East of IL-145, Golconda, IL	1:21	2:41

Southern Illinois

Scan for Map   **Start Times = Central Daylight Time**

1 Jaycee's Ball Park
800 N Moore St, Waterloo, IL 62298
38.34780 -90.15772
Totality Starts at 1:17 / Lasts 2:07

2 William Zimmer Memorial Park
730 Rogers St, Waterloo, IL 62298
38.35129 -90.14420
Totality Starts at 1:17 / Lasts 2:05

3 Konarcik Park
316 N Library St, Waterloo, IL 62298
38.34059 -90.12086
Totality Starts at 1:17 / Lasts 2:05

4 Hecker Village Park
N 2nd St, Hecker IL 62248
38.30610 -89.99600
Totality Starts at 1:18 / Lasts 2:00

5 Ratz Memorial City Park
Red Bud, IL 62278
38.21283 -89.99937
Totality Starts at 1:18 / Lasts 2:20

6 Lincoln Park
Red Bud, IL 62278
38.20926 -90.00098
Totality Starts at 1:18 / Lasts 2:20

7 Fort de Chartres
IL-155, Prairie Du Rocher, IL 62277
38.08825 -90.15981
Totality Starts at 1:17 / Lasts 2:37

8 Fort Kaskaskia
4372 Park Rd, Ellis Grove, IL 62241
37.96911 -89.91066
Totality Starts at 1:18 / Lasts 2:38

Scan for Map

9 Randolph County Lake
4301 N Lake Dr, Chester, IL 62233
37.97586 -89.80148
Totality Starts at 1:18 / Lasts 2:35

10 Pierre Menard Home Historic Site
Ellis Grove, IL 62241
37.93080 -89.84414
Totality Starts at 1:18 / Lasts 2:38

11 Seagar Memorial Park
10 Truck Bypass, Chester, IL 62233
37.90517 -89.83393
Totality Starts at 1:18 / Lasts 2:40

12 Cole Memorial Park
Chester, IL 62233
37.90333 -89.81652
Totality Starts at 1:18 / Lasts 2:39

13 Tom Reid Park
Sparta, IL 62286
38.13008 -89.70165
Totality Starts at 1:18 / Lasts 2:14

14 Brown-Stevenson City Park
Sparta, IL 62286
38.11524 -89.70687
Totality Starts at 1:18 / Lasts 2:17

15 City Park
Fairground Rd, Pinckneyville, IL 62274
38.07513 -89.39667
Totality Starts at 1:19 / Lasts 1:58

16 Duquoin State Fairgrounds
655 Executive Dr, Du Quoin, IL 62832
37.98126 -89.22605
Totality Starts at 1:20 / Lasts 2:04

17 Bower Park
Ava, IL 62907
37.88548 -89.49282
Totality Starts at 1:19 / Lasts 2:33

Plan to be at your eclipse viewing site at least two hours before totality starts.

Scan for Map

18 Dry Lake Campground
43 Dry Lake Rd, Murphysboro, IL 62966
37.83218 -89.40637
Totality Starts at 1:19 / Lasts 2:34

19 Paul Ice Recreation Area
Murphysboro, IL 62966
37.81240 -89.41372
Totality Starts at 1:19 / Lasts 2:35

20 Glenn Schlimpert Recreation Area
Murphysboro, IL 62966
37.80227 -89.39912
Totality Starts at 1:19 / Lasts 2:35

21 Lake Kinkaid Ramp
Marina Rd, Murphysboro, IL 62966
37.79748 -89.41472
Totality Starts at 1:19 / Lasts 2:36

22 Crisenberry Dam
Spillway Rd, Gorham, IL 62940
37.77796 -89.45189
Totality Starts at 1:19 / Lasts 2:38

23 Lake Murphysboro State Park
Murphysboro, IL 62966
37.77296 -89.37883
Totality Starts at 1:19 / Lasts 2:37

24 Riverside Park
Murphysboro, IL 62966
37.75693 -89.35696
Totality Starts at 1:19 / Lasts 2:37

25 CP 11462 IL-3
Gorham, IL 62940
37.72246 -89.46397
Totality Starts at 1:19 / Lasts 2:41

26 Devil's Backbone Campground
Brunkhorst Ave
Grand Tower, IL 62942
37.63963 -89.51321
Totality Starts at 1:19 / Lasts 2:40

Scan for Map

27 Oakwood Bottoms Greentree Reservoir Interpretive Site
Grand Tower, IL 62942
37.67456 -89.46210
Totality Starts at 1:19 / Lasts 2:41

28 Turkey Bayou Campground
Pomona, IL 62975
37.68507 -89.41235
Totality Starts at 1:19 / Lasts 2:41

29 Oakdale Park
Carbondale, IL 62901
37.73747 -89.23187
Totality Starts at 1:20 / Lasts 2:35

30 Parrish Park
Carbondale, IL 62901
37.72875 -89.24766
Totality Starts at 1:20 / Lasts 2:36

31 Oakland Field
Carbondale, IL 62901
37.72775 -89.23005
Totality Starts at 1:20 / Lasts 2:36

32 Doug Lee Park
Carbondale, IL 62901
37.71980 -89.19757
Totality Starts at 1:20 / Lasts 2:35

33 Carbondale Reservoir Dam
W Pleasant Hill Rd, Carbondale, IL
37.70015 -89.22387
Totality Starts at 1:20 / Lasts 2:37

34 Pomona Natural Bridge
Natural Bridge Rd, Pomona, IL 62975
37.64860 -89.34292
Totality Starts at 1:19 / Lasts 2:41

35 CP 4630 IL-127
Carbondale, IL 62903
37.66287 -89.31846
Totality Starts at 1:19 / Lasts 2:41

Plan to be at your eclipse viewing site at least two hours before totality starts.

Scan for Map

 36 Poplar Camp Beach
Shawnee National Forest, 1899 Cedar
Creek Rd, Carbondale, IL 62903
37.66641 -89.27056
Totality Starts at 1:20 / Lasts 2:40

 37 Cove Hollow
Cedar Lake Cove Hollow Rd
Carbondale, IL 62903
37.63990 -89.29111
Totality Starts at 1:20 / Lasts 2:41

 38 Pine Hills Campground
Wolf Lake, IL 62998
37.51593 -89.42235
Totality Starts at 1:19 / Lasts 2:33

 39 Trail of Tears State Forest
State Forest Rd, Jonesboro, IL 62952
37.48612 -89.36505
Totality Starts at 1:20 / Lasts 2:33

 40 Lincoln Memorial Picnic Grounds
N Main St, Jonesboro, IL 62952
37.45866 -89.26990
Totality Starts at 1:20 / Lasts 2:34

 41 Anna City Park
Anna, IL 62906
37.45566 -89.24500
Totality Starts at 1:20 / Lasts 2:35

 42 CP HWY 51
2474 S Illinois Ave, Makanda, IL 6295
37.62871 -89.23564
Totality Starts at 1:20 / Lasts 2:41

 43 CP 551 Makanda Rd
Makanda, IL 62958
37.61789 -89.20940
Totality Starts at 1:20 / Lasts 2:41

 44 Giant City State Park
842 Giant City Rd, Makanda, IL 62958
37.60797 -89.18537
Totality Starts at 1:20 / Lasts 2:41

Scan for Map

 45 Long View Park
Crab Orchard Lake, County Rd 26
Carbondale, IL 62902
37.73382 -89.14115
Totality Starts at 1:20 / Lasts 2:33

 46 Crab Orchard Campground
Campground Dr, Carterville, IL 62918
37.73907 -89.12577
Totality Starts at 1:20 / Lasts 2:33

 47 Little Grassy Lake Campground
788 Hidden Bay Ln, Makanda, IL 60058
37.64302 -89.14146
Totality Starts at 1:20 / Lasts 2:38

 48 Devils Kitchen Lake
Grassy Rd, Makanda, IL 62958
37.64400 -89.10632
Totality Starts at 1:20 / Lasts 2:38

 49 Camp Cedar Point
1327 Camp Cedar Point Ln
Makanda, IL 62958
37.62586 -89.13285
Totality Starts at 1:20 / Lasts 2:39

 50 CP Corner of Water Valley Rd &
S Rocky Comfort Rd
Makanda, IL 62958
37.57939 -89.11646
Totality Starts at 1:20 / Lasts 2:41

 51 Super-Koll Park
Maple St, Royalton, IL 62983
37.88965 -89.06122
Totality Starts at 1:20 / Lasts 2:10

 52 Four Seasons Campground
720 E Carroll St, Herrin, IL 62948
37.81489 -89.01774
Totality Starts at 1:20 / Lasts 2:19

 53 Arrowhead Lake Camp Ground
Peterson Ave, Johnston City, IL 62951
37.82841 -88.91072
Totality Starts at 1:20 / Lasts 2:10

Plan to be at your eclipse viewing site at least two hours before totality starts.

Scan for Map

54 Pigeon Creek Rec Area
Pigeon Creek Rd, Marion, IL 62959
37.72019 -89.02771
Totality Starts at 1:20 / Lasts 2:31

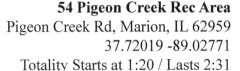

63 CP 5476 IL-37
Goreville, IL 62939
37.51624 -88.96448
Totality Starts at 1:21 / Lasts 2:41

55 Ray Fosse Park
500 E Deyoung St, Marion, IL 62959
37.74237 -88.91931
Totality Starts at 1:20 / Lasts 2:24

64 Dutchman Lake
Fishing Hole Ln, Ozark, IL 62972
37.48951 -88.91465
Totality Starts at 1:20 / Lasts 2:41

56 Williamson County Fairgrounds
399 N Fair St, Marion, IL 62959
37.73294 -88.91263
Totality Starts at 1:20 / Lasts 2:25

65 CP I-24
Vienna, IL 62995
37.48638 -88.89290
Totality Starts at 1:21 / Lasts 2:41

57 Pyramid Park
S Bevabeck Dr, Marion, IL 62959
37.71597 -88.93759
Totality Starts at 1:20 / Lasts 2:28

66 CP 4200 US-45
Vienna, IL 62995
37.47246 -88.85960
Totality Starts at 1:21 / Lasts 2:41

58 Marion Reservoir
11289 Reservoir Rd, Marion, IL 62959
37.67759 -88.95397
Totality Starts at 1:20 / Lasts 2:32

67 Tunnel Hill State Trail
302 E Vine St, Vienna, IL 62995
37.41770 -88.88816
Totality Starts at 1:21 / Lasts 2:40

59 Lake Egypt Dam
Lake Egypt Rd, Marion, IL 62959
37.62492 -88.94279
Totality Starts at 1:20 / Lasts 2:35

68 Ball Park Rd
Vienna, IL 62995
37.41364 -88.90408
Totality Starts at 1:21 / Lasts 2:40

60 Buck Ridge Campground
Ozark, IL 62972
37.58063 -88.88440
Totality Starts at 1:20 / Lasts 2:36

69 Cache River State Natural Area
930 Sunflower Ln, Belknap, IL 62908
37.37257 -88.97536
Totality Starts at 1:21 / Lasts 2:38

61 CP I-57
Buncombe, IL 62912
37.54884 -89.04284
Totality Starts at 1:20 / Lasts 2:41

70 Cache River State Natural Area
Wildcat Bluff Rd, Vienna, IL 62995
37.37671 -88.92861
Totality Starts at 1:21 / Lasts 2:39

62 Ferne Clyffe Lake
Goreville, IL 62939
37.53448 -88.97632
Totality Starts at 1:20 / Lasts 2:40

71 Mermet Swamp Nature Preserve
Belknap, IL 62908
37.25184 -88.83461
Totality Starts at 1:21 / Lasts 2:31

Plan to be at your eclipse viewing site at least two hours before totality starts.

Scan for Map

72 CP 4142 IL-147
Vienna, IL 62995
37.45296 -88.81297
Totality Starts at 1:21 / Lasts 2:41

73 Millstone Bluff National Site
Simpson, IL 62985
37.46719 -88.68791
Totality Starts at 1:21 / Lasts 2:38

74 Hayes Canyon Campground
186 Main St, Eddyville, IL 62928
37.51144 -88.58938
Totality Starts at 1:21 / Lasts 2:33

75 Lake Glendale Recreation Area
Golconda, IL 62938
37.41559 -88.66131
Totality Starts at 1:21 / Lasts 2:40

76 Duck Bay Campground
Lake Glendale, Golconda, IL 62938
37.41006 -88.65692
Totality Starts at 1:21 / Lasts 2:40

77 CP IL-145
Golconda, IL 62938
37.39374 -88.67177
Totality Starts at 1:21 / Lasts 2:41

78 Dixon Springs State Park
IL-146, Golconda, IL 62938
37.38541 -88.66831
Totality Starts at 1:21 / Lasts 2:41

79 CP 958 IL-146
Golconda, IL 62938
37.38005 -88.63916
Totality Starts at 1:21 / Lasts 2:41

80 Musketeer Campground
Metropolis, IL 62960
37.14692 -88.70867
Totality Starts at 1:21 / Lasts 2:24

Scan for Map

81 Fort Massac State Park
1308 E 5th St, Metropolis, IL 62960
37.14289 -88.71006
Totality Starts at 1:21 / Lasts 2:23

82 Garden of the Gods Rec Area
Picnic Rd, Herod, IL 62947
37.60514 -88.38456
Totality Starts at 1:22 / Lasts 2:11

83 Pharaoh Campground
Herod, IL 62947
37.60239 -88.37895
Totality Starts at 1:22 / Lasts 2:11

84 Cave-In-Rock State Park
New State Park Rd, Cave-In-Rock, IL
37.46975 -88.15527
Totality Starts at 1:22 / Lasts 2:17

85 Ohio River
Main St, Rosiclare, IL 62982
37.41799 -88.34368
Totality Starts at 1:22 / Lasts 2:32

86 Golconda Historic District
National Register Site
Golconda, IL 62938
37.34658 -88.49216
Totality Starts at 1:22 / Lasts 2:40

87 Deer Run Campground
RR 3, Golconda, IL 62938
37.33818 -88.50007
Totality Starts at 1:22 / Lasts 2:40

88 CP Ohio River Scenic Byway
Golconda, IL 62938
37.32397 -88.50615
Totality Starts at 1:22 / Lasts 2:41

89 Ohio River at Bay City
Bay City Rd, Golconda, IL 62938
37.24962 -88.49654
Totality Starts at 1:22 / Lasts 2:40

Plan to be at your eclipse viewing site at least two hours before totality starts.

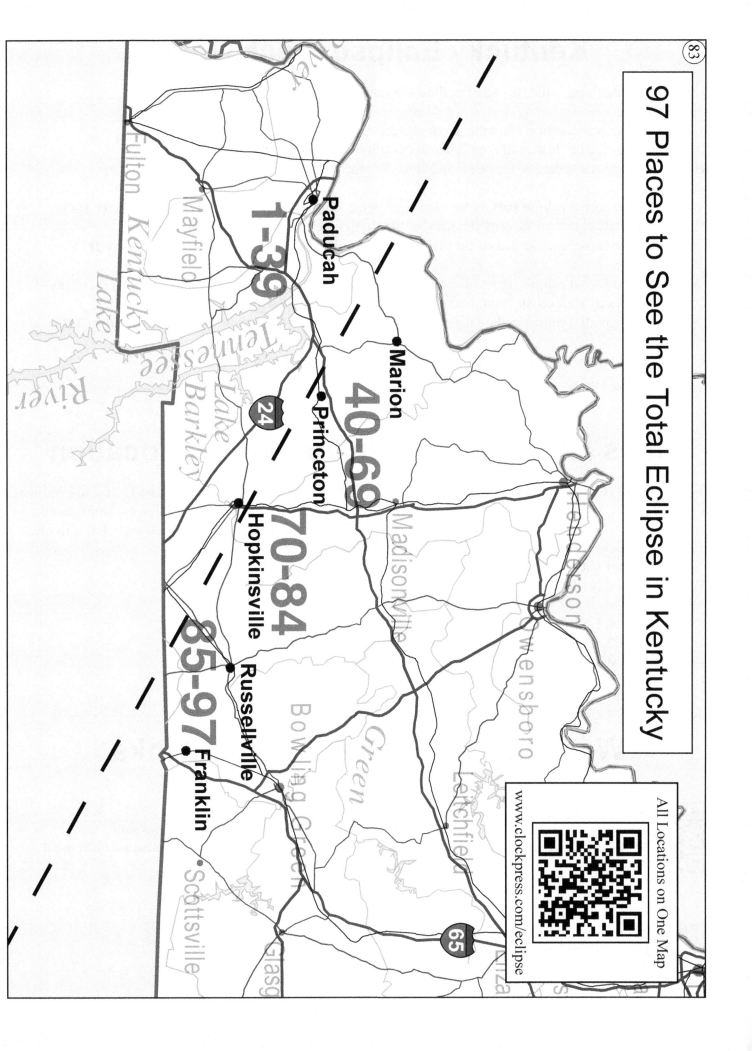

97 Places to See the Total Eclipse in Kentucky

All Locations on One Map

www.clockpress.com/eclipse

Kentucky Eclipse Track Notes

How many miles long is the eclipse centerline in Kentucky? **98 miles**

What is the average duration of totality along the centerline in Kentucky? **2:41 (m:ss)**
How long does totality last on the centerline at the Illinois/Kentucky border? **2:41 (m:ss)**
How long does totality last on the centerline at the Kentucky/Tennessee border? **2:41 (m:ss)**

When does the partial eclipse start on the centerline at the Illinois/Kentucky border? **11:54 a.m. (CDT)**
When does totality start on the centerline at the Illinois/Kentucky border? **1:22 p.m. (CDT)**
When does the partial eclipse end on the centerline at the Illinois/Kentucky border? **2:49 p.m. (CDT)**

When does the partial eclipse start on the centerline at the Kentucky/Tennessee border? **11:57 a.m. (CDT)**
When does totality start on the centerline at the Kentucky/Tennessee border? **1:26 p.m. (CDT)**
When does the partial eclipse end on the centerline at the Kentucky/Tennessee border? **2:52 p.m. (CDT)**

Check local forecasts and satellite maps for the best weather information for this eclipse.

Cities and Highways in Totality by Location

1-12: Ohio River - **I-24, US-60** - Carrsville, Smithland, Paducah

13-44: Land Between the Lakes - **I-24, US-68, US-62** Benton, Calvert City, Grand Rivers, Aurora, Golden Pond, Cadiz, Eddyville

45-70: Between I-69 & I-24 - **I-69, US-41, US-641, US-431, US-62, Pennyrile Parkway** - Dycusburg, Fredonia, Marion, Providence, Earlington, Nortonville, Dawson Springs, Princeton, Cerulean

71-97: East of I-24 & West of I-65 - **I-24, US-68, US-41, US-431, US-79, I-65** - Gracey, Hopkinsville, Ft. Campbell, Belton, Lewisburg, Russellville, Franklin, Adairville

Weather and Eclipse Related Links

Website

Kentucky Weather
Paducah, KY NWS Office
http://www.weather.gov/pah/

Hopkinsville Kentucky Eclipse Site
http://www.visithopkinsville.com/listing/solar-eclipse-2017/

Midwest U.S. GOES Satellite - Visible Loop
http://www.ssd.noaa.gov/goes/east/mw/h5-mloop-vis.html

Featured Eclipse Destinations

Kentucky Dam Village State Park has many attractions including lodging, camping and a huge marina. Eclipse watchers will see nearly two and a half minutes of totality from this park. Totality starts at 1:22 p.m. CDT. See Kentucky #17 for location details.

Website

Photos

Pennyrile Forest State Park has kayaking, horseback riding, hiking trails and many other attractions. During the eclipse the totality phase will last two minutes and thirty five seconds from this location. Totality starts at 1:24 p.m. CDT. See Kentucky #63 for details.

Website

Photos

Lake Barkley State Park is full featured with boating, camping, swimming, bike trails and many more attractions. Visitors will experience two and a half minutes of totality during the eclipse. Totality starts at 1:23 p.m. CDT. See Kentucky #37 for details.

Website

Photos

Jeffers Bend Recreational Area consists of walking trails, a small lake and a botanical garden. It is situated right on the centerline of the eclipse and will see two minutes and forty one seconds of totality. Totality starts at 1:24 p.m. CDT. See Kentucky #77 for location details.

Website

Photos

KY Crossing Points Table

Where the Eclipse Centerline Crosses Highways in Kentucky

Start Times = Central Daylight Time

Loc#	Hwy	Nearest Mile Marker, Cross Street or Exit, City	TStart	TLasts
2	US-60	MM 22, Burna, KY	1:22	2:41
45	US-641	KY-1943, Eddyville, KY	1:23	2:41
56	I-69	MM 6, Pleasant Valley Rd, Princeton, KY	1:23	2:41
70	KY-624	J Stewart Cemetery Rd, Cerulean, KY	1:24	2:41
72	KY-91	KY-1026, Cerulean, KY	1:24	2:41
76	US-41	Talbert Dr, Hopkinsville, KY	1:24	2:41
78	Pennyrile Pkwy	MM 11, Exit 11, Hopkinsville, KY	1:24	2:41
87	US-68	Edwards Mill Rd, Hopkinsville, KY	1:24	2:41
91	US-79	.8 miles South of KY-102, Guthrie, KY	1:25	2:41

Western Kentucky

Start Times = Central Daylight Time

1 Ohio River at Carrsville
2127 Main St, Carrsville, KY 42081
37.39935 -88.37259
Totality Starts at 1:22 / Lasts 2:34

2 CP 1744 US-60
Burna, KY 42028
37.24951 -88.33036
Totality Starts at 1:22 / Lasts 2:41

3 Livingston County Fairgrounds
799 U 60 E, Smithland, KY 42081
37.15414 -88.39574
Totality Starts at 1:22 / Lasts 2:39

4 Ohio River at Smithland
371 Riverfront Dr
Smithland, KY 42081
37.14276 -88.40595
Totality Starts at 1:22 / Lasts 2:39

5 Lone Oak Park
Denver Ave, Paducah, KY 42001
37.03888 -88.67496
Totality Starts at 1:22 / Lasts 2:08

6 Stuart Nelson Park
S Nelson Park Rd, Paducah, KY 42001
37.09094 -88.65618
Totality Starts at 1:22 / Lasts 2:18

7 Noble Park
Park Ave, Paducah, KY 42001
37.08738 -88.63980
Totality Starts at 1:22 / Lasts 2:19

8 Carson Park
Harrison St, Paducah, KY 42001
37.07472 -88.63517
Totality Starts at 1:22 / Lasts 2:17

9 Albert Jones Park
Joe Bryan Dr, Paducah, KY 42003
37.07131 -88.61397
Totality Starts at 1:22 / Lasts 2:18

10 Dolly McNutt Memorial Plaza
300 S 5th St, Paducah, KY 42003
37.08332 -88.59824
Totality Starts at 1:22 / Lasts 2:21

11 Duck Creek RV Park
2540 John L. Puryear Dr, Paducah, KY
37.02134 -88.58612
Totality Starts at 1:22 / Lasts 2:11

12 CSA Memorial
6899 Reidland Rd, Paducah, KY 42003
36.99376 -88.51181
Totality Starts at 1:22 / Lasts 2:12

13 Clarks River Wildlife Refuge
91 US-641, Benton, KY 42025
36.87998 -88.34458
Totality Starts at 1:23 / Lasts 2:05

14 Mike Miller County Park
US-68, Benton, KY 42025
36.93590 -88.34897
Totality Starts at 1:23 / Lasts 2:14

15 Kentucky Lake Motor Speedway
957 Truck Plaza Ln, Calvert City, KY
36.98987 -88.33924
Totality Starts at 1:22 / Lasts 2:24

16 Cypress Lakes RV Park
54 Scillion Dr, Calvert City, KY 42029
37.00540 -88.33251
Totality Starts at 1:22 / Lasts 2:27

17 Kentucky Dam Village State Park
113 Administration Dr
Gilbertsville, KY 42044
36.99891 -88.29616
Totality Starts at 1:22 / Lasts 2:28

Plan to be at your eclipse viewing site at least two hours before totality starts.

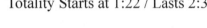

18 Kentucky Dam State Park Airport
Gilbertsville, KY 42044
37.00905 -88.29300
Totality Starts at 1:22 / Lasts 2:30

27 Hillman Ferry Campground
820 Hillman Ferry Road
Grand Rivers, KY 42045
36.94657 -88.18148
Totality Starts at 1:23 / Lasts 2:28

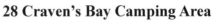

19 Kentucky Dam Campground
Gilbertsville, KY 42044
37.01333 -88.28161
Totality Starts at 1:22 / Lasts 2:31

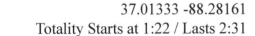

28 Craven's Bay Camping Area
Eddyville, KY 42038
36.96107 -88.05366
Totality Starts at 1:23 / Lasts 2:36

20 Crockett Frontiers Campground
Maxwell Dr, Grand Rivers, KY 42045
37.00568 -88.23967
Totality Starts at 1:23 / Lasts 2:32

29 Birmingham Ferry Rec Area
FS Road 115, Benton, KY 42025
36.92312 -88.16264
Totality Starts at 1:23 / Lasts 2:25

21 Kentucky Lake at Barkley Canal
FS Road 101, Grand Rivers, KY 42045
36.98380 -88.22601
Totality Starts at 1:23 / Lasts 2:30

30 Aurora Oaks Campground
55 KOA Rd, Kentucky 42048
36.77805 -88.14556
Totality Starts at 1:23 / Lasts 2:02

22 Canal Public Use Area
Grand Rivers, KY 42045
36.99607 -88.21184
Totality Starts at 1:23 / Lasts 2:32

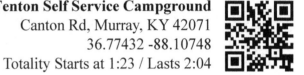

31 Fenton Self Service Campground
Canton Rd, Murray, KY 42071
36.77432 -88.10748
Totality Starts at 1:23 / Lasts 2:04

23 Cumberland River
FS Rd 102-C, Grand Rivers, KY 42045
36.98888 -88.20050
Totality Starts at 1:23 / Lasts 2:32

32 Jenny Ridge Picnic Area
The Trace Rd, Benton, KY 42025
36.78834 -88.06752
Totality Starts at 1:23 / Lasts 2:09

24 North Station Between the Lakes
1820 The Trace, Grand Rivers, KY
36.97117 -88.19910
Totality Starts at 1:23 / Lasts 2:30

33 Golden Pond
Visitor Center & Planetarium
Visitor Center Dr, Golden Pond, KY
36.77844 -88.06324
Totality Starts at 1:23 / Lasts 2:08

25 Demumbers Bay Backcountry Area
108 Kuttawa, KY 42055
36.97756 -88.15303
Totality Starts at 1:23 / Lasts 2:33

34 Linton Public Use Area
Cadiz, KY 42211
36.68769 -87.91056
Totality Starts at 1:24 / Lasts 2:04

26 Moss Creek Day Use Area
Gilbertsville, KY 42044
36.95071 -88.19627
Totality Starts at 1:23 / Lasts 2:27

35 Cadiz Park
Jefferson St, Cadiz, KY 42211
36.85937 -87.84116
Totality Starts at 1:24 / Lasts 2:34

Plan to be at your eclipse viewing site at least two hours before totality starts.

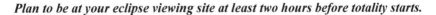

Scan for Map   Scan for Map

36 Kamptown RV Resort
4124 Rockcastle Rd, Cadiz, KY 42211
36.87884 -87.90329
Totality Starts at 1:23 / Lasts 2:34

45 CP 2798 US-641
Eddyville, KY 42038
37.13682 -88.06604
Totality Starts at 1:23 / Lasts 2:41

37 Lake Barkley State Park
3500 State Park Rd, Cadiz, KY 42211
36.84844 -87.91419
Totality Starts at 1:23 / Lasts 2:30

46 Cumberland River at Dycusburg
Commerce St, Marion, KY 42064
37.15971 -88.18688
Totality Starts at 1:22 / Lasts 2:41

38 Lake Barkley State Park Airport
Cadiz, KY 42211
36.82377 -87.90465
Totality Starts at 1:24 / Lasts 2:27

47 Maple Lake
327 Main Lake Rd, Marion, KY 42064
37.18362 -88.15655
Totality Starts at 1:22 / Lasts 2:41

39 Kentucky Lakes / Prizer KOA
1777 Prizer Point Rd, Cadiz, KY 42211
36.91181 -87.99027
Totality Starts at 1:23 / Lasts 2:33

48 Marion-Crittenden Airport
449 Airport Rd, Marion, KY 42064
37.33669 -88.10863
Totality Starts at 1:22 / Lasts 2:31

40 Eddy Creek Marina Resort
7612 KY-93, Eddyville, KY 42038
37.00718 -88.01822
Totality Starts at 1:23 / Lasts 2:39

49 War Eagle RV Park
1597 US-60, Marion, KY
37.32980 -88.10526
Totality Starts at 1:22 / Lasts 2:32

41 Indian Point RV Park
1136 Indian Hills Trail, Eddyville, KY
37.03555 -88.05575
Totality Starts at 1:23 / Lasts 2:40

50 Veterans Memorial Park
N Weldon St, Marion, KY 42064
37.33748 -88.08499
Totality Starts at 1:22 / Lasts 2:30

42 Mineral Mound State Park
48 Finch Ln, Eddyville, KY 42038
37.06048 -88.08624
Totality Starts at 1:23 / Lasts 2:40

51 Providence City Lake
Providence, KY 42450
37.37621 -87.79720
Totality Starts at 1:23 / Lasts 2:06

43 Vista Ridge Park
Kuttawa, KY 42055
37.06583 -88.11698
Totality Starts at 1:23 / Lasts 2:40

52 Cube Park
Rufus Rd, Princeton, KY 42445
37.23626 -87.88038
Totality Starts at 1:23 / Lasts 2:31

44 Venture River Water Park
280 Park Pl, Eddyville, KY 42038
37.08754 -88.09373
Totality Starts at 1:23 / Lasts 2:40

53 "Sis" Baker Park
Marion Rd, Fredonia, KY 42411
37.20865 -88.05342
Totality Starts at 1:23 / Lasts 2:37

Plan to be at your eclipse viewing site at least two hours before totality starts.

Scan for Map

54 Paul F. Riley W.O.W Park
Grove St, Fredonia, KY 42411
37.20476 -88.05755
Totality Starts at 1:23 / Lasts 2:37

55 Buddy Rogers Park
E Crider St, Fredonia, KY 42411
37.20396 -88.05822
Totality Starts at 1:23 / Lasts 2:37

56 CP Western KY Parkway
KY-9001 Princeton, KY 42445
37.10817 -87.99916
Totality Starts at 1:23 / Lasts 2:41

57 Ratliff Park
Princeton, KY 42445
37.10029 -87.88574
Totality Starts at 1:23 / Lasts 2:39

58 CP 4992 US-62
Princeton, KY 42445
37.10444 -87.99049
Totality Starts at 1:23 / Lasts 2:41

59 VFW Park
Princeton, KY 42445
37.09861 -87.88339
Totality Starts at 1:23 / Lasts 2:39

60 Railroad Lake
Princeton, KY 42445
37.08026 -87.86234
Totality Starts at 1:23 / Lasts 2:40

61 Dawson Springs City Park
Dawson Springs, KY 42408
37.17243 -87.68618
Totality Starts at 1:23 / Lasts 2:30

62 Lake Beshear
Dawson Springs, KY 42408
37.14103 -87.68040
Totality Starts at 1:24 / Lasts 2:32

Scan for Map

63 Pennyrile Forest State Park
Dawson Springs, KY 42408
37.07323 -87.66290
Totality Starts at 1:24 / Lasts 2:35

64 Earlington City Park
Earlington Rd, Earlington, KY 42410
37.27395 -87.52382
Totality Starts at 1:24 / Lasts 2:01

65 Nortonville Lake
High St, Nortonville, KY 42442
37.17988 -87.46412
Totality Starts at 1:24 / Lasts 2:16

66 White Plains City Park
299 S Bob Bruce Dr
White Plains, KY 42464
37.18418 -87.38381
Totality Starts at 1:24 / Lasts 2:09

67 Gordon Park
13708 N Old Madisonville Rd
Crofton, Kentucky
37.05829 -87.48693
Totality Starts at 1:24 / Lasts 2:32

68 Lake Malone State Park
Belton, KY 42324
37.06997 -87.04830
Totality Starts at 1:25 / Lasts 2:00

69 Lewisburg City Park
Woodlawn St, Lewisburg, KY 42256
36.99175 -86.94503
Totality Starts at 1:25 / Lasts 2:09

70 CP Cerulean
Hopkinsville Rd, Near Quarry Rd,
Cerulean, KY 42215
36.96590 -87.66908
Totality Starts at 1:24 / Lasts 2:41

71 Gracey Christian Park
E Main St, Gracey, KY 42232
36.87683 -87.65844
Totality Starts at 1:24 / Lasts 2:40

Plan to be at your eclipse viewing site at least two hours before totality starts.

Scan for Map

72 CP 8699 Princeton Rd
Cerulean, KY 42215
36.93847 -87.60582
Totality Starts at 1:24 / Lasts 2:41

73 Lake Morris
1698 Morris Lake Rd
Hopkinsville, KY 42240
36.92812 -87.45691
Totality Starts at 1:24 / Lasts 2:38

74 Ruff Park
Litchfield Dr, Hopkinsville, KY 42240
36.88497 -87.49307
Totality Starts at 1:24 / Lasts 2:41

75 University of Kentucky Community College
Hopkinsville, KY 42240
36.88307 -87.48960
Totality Starts at 1:24 / Lasts 2:41

76 CP 1024 N Main St
Hopkinsville, KY 42240
36.88456 -87.48177
Totality Starts at 1:24 / Lasts 2:41

77 Jeffers Bend Recreation Area
Hopkinsville, KY 42240
36.87983 -87.47089
Totality Starts at 1:24 / Lasts 2:41

78 CP Pennyrile Pkwy
Hopkinsville, KY 42240
36.87874 -87.46843
Totality Starts at 1:24 / Lasts 2:41

79 West 1st Street Park
101-199 River Front St
Hopkinsville, KY 42240
36.87123 -87.48874
Totality Starts at 1:24 / Lasts 2:41

80 Little River Park
Bethel St, Hopkinsville, KY 42240
36.86657 -87.49009
Totality Starts at 1:24 / Lasts 2:41

Scan for Map

81 Peace Park
700-798 S Campbell St
Hopkinsville, KY 42240
36.86337 -87.48468
Totality Starts at 1:24 / Lasts 2:41

82 Cherokee Trail Of Tears Commemorative Park
100 E 9th St, Hopkinsville, KY 42240
36.85353 -87.46968
Totality Starts at 1:24 / Lasts 2:41

83 Christian County Airport
Memorial Field Dr, Hopkinsville, KY
36.85712 -87.45506
Totality Starts at 1:24 / Lasts 2:41

84 Fort Campbell Memorial Park
Hopkinsville, KY 42240
36.83247 -87.47146
Totality Starts at 1:24 / Lasts 2:40

85 James Bruce Convention Center
303 Conference Center Dr
Hopkinsville, KY 42240
36.80633 -87.48111
Totality Starts at 1:24 / Lasts 2:40

86 Mc Dagin Park
Fort Campbell, KY 42223
36.65930 -87.45420
Totality Starts at 1:25 / Lasts 2:31

87 CP 4979 US-68
Jefferson Davis Hwy
Hopkinsville, KY 42240
36.85491 -87.41373
Totality Starts at 1:24 / Lasts 2:41

88 Jefferson Davis State Historic Site
258 Pembroke-Fairview Rd
Pembroke, KY 42266
36.84240 -87.29988
Totality Starts at 1:25 / Lasts 2:39

89 Bradley-Crusher Park
313 Clarksville Rd, Trenton, KY 42286
36.72099 -87.26366
Totality Starts at 1:25 / Lasts 2:40

Plan to be at your eclipse viewing site at least two hours before totality starts.

Scan for Map

90 Public Square
Elkton, KY 42220
36.80974 -87.15405
Totality Starts at 1:25 / Lasts 2:37

91 CP 2853 US-79
Russellville Rd, Guthrie, KY 42234
36.71574 -87.09618
Totality Starts at 1:25 / Lasts 2:41

92 Logan County Memorial Park
Bobby Sawyer Way, Russellville, KY
36.85978 -86.91691
Totality Starts at 1:25 / Lasts 2:26

93 Community Park
Adairville, KY 42202
36.67441 -86.85520
Totality Starts at 1:26 / Lasts 2:37

94 Auburn Park
Spring St, Auburn, KY 42206
36.86106 -86.70585
Totality Starts at 1:26 / Lasts 2:13

95 Jim Roberts Community Park
North St, Franklin, KY 42134
36.73089 -86.54689
Totality Starts at 1:26 / Lasts 2:22

96 I-65 Rest Area
Franklin, KY 42134
36.64373 -86.56908
Totality Starts at 1:26 / Lasts 2:32

97 Kentucky Downs
5629 Nashville Rd, Franklin, KY 42134
36.64212 -86.56351
Totality Starts at 1:26 / Lasts 2:32

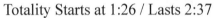
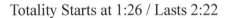
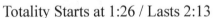

Plan to be at your eclipse viewing site at least two hours before totality starts.

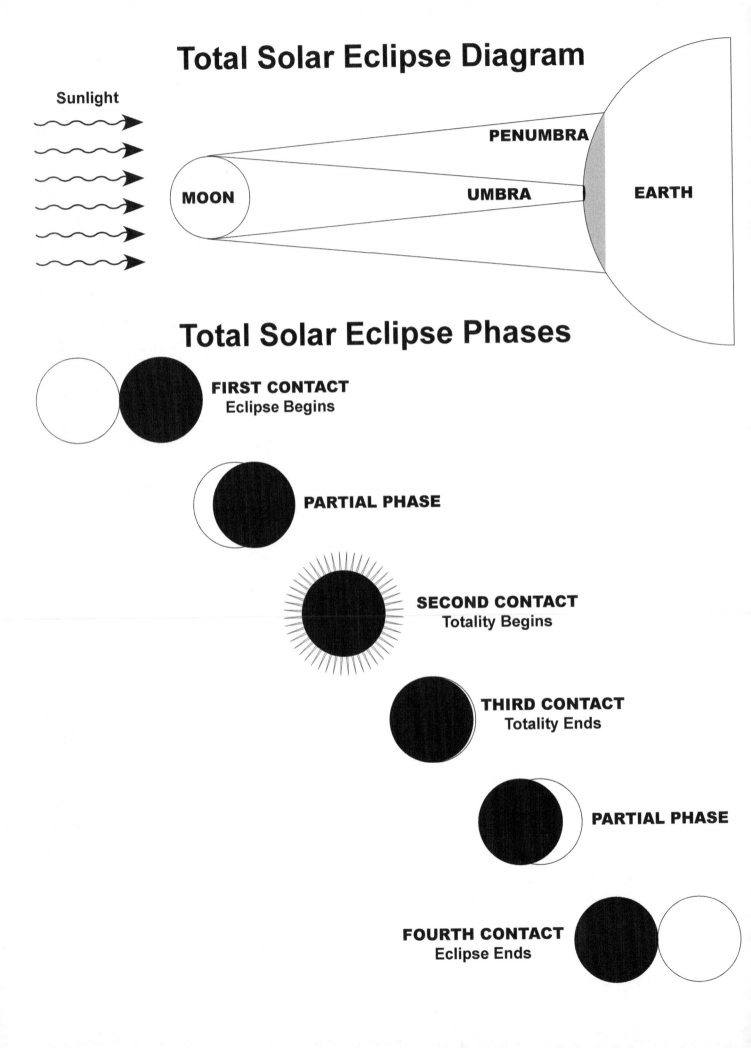

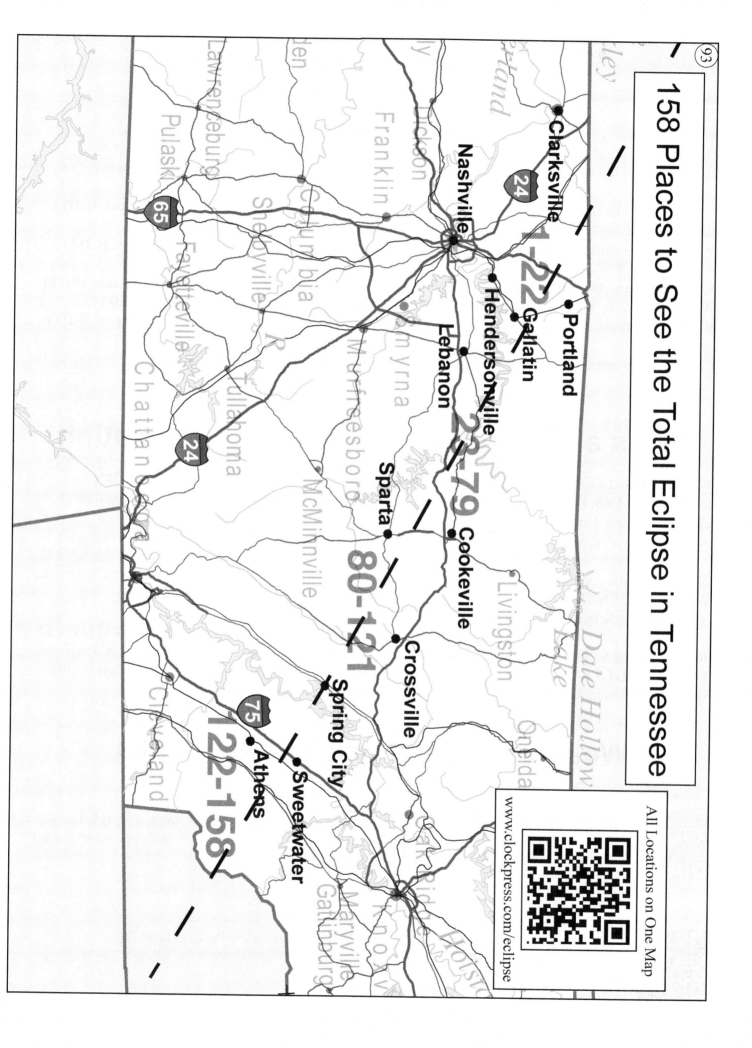

158 Places to See the Total Eclipse in Tennessee

All Locations on One Map

www.clockpress.com/eclipse

1-22

23-79

80-121

122-158

Clarksville

Nashville

Portland

Gallatin

Hendersonville

Lebanon

Sparta

Cookeville

Crossville

Spring City

Athens

Sweetwater

Tennessee Eclipse Track Notes

How many miles long is the eclipse centerline in Tennessee? **98 miles**

What is the average duration of totality along the centerline in Tennessee? **2:40 (m:ss)**
How long does totality last on the centerline at the Kentucky/Tennessee border? **2:41 (m:ss)**
How long does totality last on the centerline at the Tennessee/North Carolina border? **2:39 (m:ss)**

When does the partial eclipse start on the centerline at the Kentucky/Tennessee border? **11:57 a.m. (CDT)**
When does totality start on the centerline at the Kentucky/Tennessee border? **1:26 p.m. (CDT)**
When does the partial eclipse end on the centerline at the Kentucky/Tennessee border? **2:52 p.m. (CDT)**

When does the partial eclipse start on the centerline at the Tennessee/North Carolina border? **1:05 p.m. (EDT)**
When does totality start on the centerline at the Tennessee/North Carolina border? **2:33 p.m. (EDT)**
When does the partial eclipse end on the centerline at the Tennessee/North Carolina border? **3:59 p.m. (EDT)**

Check local forecasts and satellite maps for the best weather information for this eclipse.

Cities and Highways in Totality by Location

1-22: North and East of Nashville - **US-79, US-41, I-24, US-431, I-65, US-31W, TN-109, TN-52** Clarksville, Port Royal, Springfield, Cross Plains, Portland, Lafayette, White House, Goodlettsville

23-39: Nashville and East - **US-41, US-431, I-24, TN-155, I-65, US-31E, TN-386** Nashville, Goodlettsville, Hendersonville

40-69: East and South of Nashville - **US-41, I-40, TN-840, US-231, US-70, TN-386, TN-25** - Hermitage, Mt. Juliet, Lebanon, Gallatin, Castalian Springs

70-86: I-40 Corridor - **US-70, I-40, US-70N** Watertown, Alexandria, Carthage, Gordonsville, Silver Point, Lancaster

87-121: I-40 Corridor - **I-40, US-70, TN-56, US-127, US-70S** - Smithville, Sparta, Rock Island, Cookeville, Monterey, Crossville, Spencer, Pikeville

122-158: Eastern Time Zone - **I-40, US-127, US-27, I-75, US-11, US-411, US-129** - Spring City, Harriman, Kingston, Decatur, Lenoir City, Sweetwater, Athens, Madisonville, Vonore, Tellico Plains

Weather and Eclipse Related Links

Website

Middle Tennessee Weather
Nashville, TN NWS Office
http://www.srh.noaa.gov/ohx/

Midwest U.S. GOES Satellite - Visible Loop
http://www.ssd.noaa.gov/goes/east/mw/h5-mloop-vis.html

Eastern Tennessee Weather
Morristown, TN NWS Office
http://www.srh.noaa.gov/mrx/

Tennessee State Parks Eclipse Site
http://tnstateparks.com/activities/solar-eclipse-at-the-park-2017

Dunbar Cave State Park features several hiking trails, a picnic area and fishing as well as the historic Dunbar Cave. Check out the cave after the eclipse and see two minutes and twenty five seconds of totality on eclipse day. Totality starts at 1:25 p.m. CDT. See Tennessee #4 for location details.

Website

Photos

Edgar Evins State Park on Center Hill Lake is over 6,000 acres in size. Cabins, camping, fishing, hiking and many other attractions are offered. Eclipse watchers can see two minutes and thirty nine seconds of totality from this state park. Totality starts at 1:29 p.m. CDT. See Tennessee #86 for details.

Website

Photos

Cedars of Lebanon State Park is over 1,000 acres and has many attractions including horse stables, hiking, a swimming pool, picnic areas and many more features. Eclipse day visitors will see two minutes and twenty seconds of totality from this site. Totality starts at 1:28 p.m. CDT. See Tennessee #49 for details.

Website

Photos

Rock Island State Park is over 800 acres in size and features scenic overlooks, picnic shelters, a beach and campgrounds. Visitors on eclipse day will see two minutes and twenty six seconds of totality. Totality starts at 1:29 p.m. CDT. See Tennessee #92 for location details.

Website

Photos

TN Crossing Points Table

Where the Eclipse Centerline Crosses Highways in Tennessee				
		C = Central Daylight Time E = Eastern Daylight Time		
Loc#	Hwy	Nearest Mile Marker, Cross Street or Exit, City	TStart	TLasts
8	US-431/TN-65	Borthick Rd, Springfield, TN	1:26C	2:41
19	I-65	MM 112, Main St, Cross Plains, TN	1:26C	2:40
56	US-31E	Airport Rd, Gallatin, TN	1:27C	2:40
61	US-231/TN-10	Crook Ln, Castalian Springs, TN	1:27C	2:40
72	US-70N	Short St, Rome Access Area, Lebanon, TN	1:28C	2:40
80	I-40	MM 267, Rest Area, Lancaster, TN	1:29C	2:40
93	I-40	MM 274, Exit 273, TN-56, Silver Point, TN	1:29C	2:40
104	US-70S	MM 13, O'Connor Rd, Sparta, TN	1:29C	2:40
107	US-70	Oak Lake Dr, Sparta, TN	1:30C	2:40
121	US-127	Browns Gap Rd, Pikeville, TN	1:31C	2:40
128	US-27	MM 23, Galloway Memorial Bridge, Spring City, TN	2:31E	2:39
137	I-75	Exit 56, Niota, TN	2:32E	2:39
138	US-11	Co Rd 351, Niota, TN	2:32E	2:39
144	US-411	Co Rd 839, Maxwell Rd, Madisonville, TN	2:32E	2:39

Central Tennessee

9 Springfield-Robertson Airport
4432 Airport Rd, Cedar Hill, TN 37032
36.53861 -86.91979
Totality Starts at 1:26 / Lasts 2:39

Scan for Map   Loc# 1 - 121 Start times are Central Daylight Time

1 Smith Branch Recreation Area
Smith Branch Rd, Clarksville, TN 37042
36.48698 -87.43639
Totality Starts at 1:25 / Lasts 2:05

10 Robertson County Fairgrounds
Springfield, TN 37172
36.52106 -86.88126
Totality Starts at 1:26 / Lasts 2:39

2 Heritage Park
1241 Peachers Mill Rd
Clarksville, TN 37042
36.58231 -87.38553
Totality Starts at 1:25 / Lasts 2:24

11 J. Travis Price Park Band Stage
Springfield, TN 37172
36.52298 -86.86129
Totality Starts at 1:26 / Lasts 2:39

3 Kings Bluff
919 Max Ct, Clarksville, TN 37043
36.50126 -87.32298
Totality Starts at 1:25 / Lasts 2:16

12 Martin Luther King Jr. Park
2623 S Main St, Springfield, TN 37172
36.47815 -86.89604
Totality Starts at 1:26 / Lasts 2:38

4 Dunbar Cave State Park
401 Old Dunbar Cave Rd
Clarksville, TN 37043
36.55062 -87.30619
Totality Starts at 1:25 / Lasts 2:25

13 Orlinda City Park
7475 TN-52, Orlinda, TN 37141
36.59859 -86.71172
Totality Starts at 1:26 / Lasts 2:38

5 Civitan Park
650 Bellamy Ln, Clarksville, TN 37043
36.57216 -87.27846
Totality Starts at 1:25 / Lasts 2:30

14 Kilgore Park
Cross Plains, TN 37049
36.54770 -86.71059
Totality Starts at 1:26 / Lasts 2:40

6 Rotary Park
2308 Rotary Park Dr
Clarksville, TN 37043
36.50153 -87.27048
Totality Starts at 1:25 / Lasts 2:20

15 Portland City Park
Portland, TN 37148
36.57894 -86.51829
Totality Starts at 1:27 / Lasts 2:34

7 Port Royal State Park
3300 Old Clarksville Springfield Rd
Adams, TN 37010
36.55390 -87.14291
Totality Starts at 1:25 / Lasts 2:34

16 Meadowbrook Park
Portland, TN 37148
36.56870 -86.51426
Totality Starts at 1:27 / Lasts 2:35

8 CP US-431
Springfield, TN 37172
36.60833 -86.85306
Totality Starts at 1:26 / Lasts 2:41

17 Portland Municipal Airport
Portland, TN 37148
36.58636 -86.47854
Totality Starts at 1:27 / Lasts 2:33

Plan to be at your eclipse viewing site at least two hours before totality starts.

18 Macon County Park
407 Days Rd, Lafayette, TN 37083
36.51525 -86.01809
Totality Starts at 1:28 / Lasts 2:19

Scan for Map

19 CP I-65
Cross Plains, TN 37049
36.51758 -86.64888
Totality Starts at 1:26 / Lasts 2:40

20 White House City Park
White House, TN 37188
36.47007 -86.66272
Totality Starts at 1:26 / Lasts 2:40

21 Owl's Roost Campground
7267 Bethel Rd
Goodlettsville, TN 37072
36.41794 -86.72163
Totality Starts at 1:26 / Lasts 2:38

22 Ridgetop Station Park
1954 Woodruff Ave
Greenbrier, TN 37073
36.40452 -86.77038
Totality Starts at 1:26 / Lasts 2:36

23 Marrowbone Lake
6399 Marrowbone Lake Rd
Nashville, TN 37080
36.30397 -86.91693
Totality Starts at 1:26 / Lasts 2:13

24 Joseph Brown Mullins Park
4297 Drakewood Ln
Nashville, TN 37218
36.20806 -86.85139
Totality Starts at 1:27 / Lasts 2:00

25 Ted Rhodes Park
Mainstream Dr, Nashville, TN 37228
36.19348 -86.81724
Totality Starts at 1:27 / Lasts 2:01

26 Morgan Park
411 Hume St, Nashville, TN 37208
36.18126 -86.78969
Totality Starts at 1:27 / Lasts 2:01

27 Nissan Stadium
1 Titans Way, Nashville, TN 37213
36.16871 -86.76880
Totality Starts at 1:27 / Lasts 2:00

Scan for Map

28 Douglas Park
210 N 7th St, Nashville, TN 37206
36.17825 -86.76247
Totality Starts at 1:27 / Lasts 2:03

29 McFerrin Park
303 Berry St, Nashville, TN 37207
36.18217 -86.76596
Totality Starts at 1:27 / Lasts 2:03

30 Cleveland Park
Vernon Winfrey Avenue
Nashville, TN 37207
36.18810 -86.76007
Totality Starts at 1:27 / Lasts 2:05

31 Parkwood Park
3220 Vailview Dr, Nashville, TN 37207
36.23498 -86.77190
Totality Starts at 1:27 / Lasts 2:12

32 Cedar Hill Lake
1017 Nesbitt Dr, Nashville, TN 37207
36.27045 -86.74521
Totality Starts at 1:27 / Lasts 2:20

33 Two Rivers Campground
2616 Music Valley Dr
Nashville, TN 37214
36.23361 -86.70409
Totality Starts at 1:27 / Lasts 2:17

34 Heartland Park
Stones River Greenway
Nashville, TN 37214
36.18494 -86.66563
Totality Starts at 1:27 / Lasts 2:11

35 Peay Park
200 Memorial Dr, Goodlettsville, TN
36.32103 -86.71811
Totality Starts at 1:27 / Lasts 2:29

Plan to be at your eclipse viewing site at least two hours before totality starts.

Scan for Map   Scan for Map

36 Moss-Wright Park
745 Caldwell Dr, Goodlettsville, TN
36.31983 -86.68402
Totality Starts at 1:27 / Lasts 2:31

45 Bryant Grove Recreation Area
Barnett Rd, Mt Juliet, TN 37122
36.07333 -86.52261
Totality Starts at 1:28 / Lasts 2:03

37 Rockland Recreation Area
Hendersonville, TN 37075
36.30017 -86.64760
Totality Starts at 1:27 / Lasts 2:30

46 Poole Knobs Recreation Area
La Vergne, TN 37086
36.05580 -86.51345
Totality Starts at 1:28 / Lasts 2:00

38 Old Lock Three Access Area
Hendersonville, TN 37075
36.27707 -86.63126
Totality Starts at 1:27 / Lasts 2:29

47 Fate Sanders Recreation Area
Mt Juliet, TN 37122
36.04292 -86.48312
Totality Starts at 1:28 / Lasts 2:00

39 Shutes Branch Access Area
Old Hickory, TN 37138
36.24933 -86.56842
Totality Starts at 1:27 / Lasts 2:29

48 Nashville Superspeedway
4847-F McCrary Rd, Lebanon, TN 37090
36.04644 -86.41141
Totality Starts at 1:28 / Lasts 2:07

40 Nashville Shores
4001 Bell Rd, Hermitage, TN 37076
36.15778 -86.60499
Totality Starts at 1:27 / Lasts 2:11

49 Cedars of Lebanon State Park
328 Cedar Forest Rd, Lebanon, TN
36.07922 -86.31678
Totality Starts at 1:28 / Lasts 2:20

41 Cook Public Use Area
Old Hickory Blvd, Nashville, TN 37076
36.13174 -86.59519
Totality Starts at 1:27 / Lasts 2:08

50 Garrett Baseball Field
1480 Nashville Pike, Gallatin, TN 37066
36.36565 -86.49596
Totality Starts at 1:27 / Lasts 2:39

42 Seven Points Campground
1810 Stewarts Ferry Pike
Hermitage, TN 37076
36.13403 -86.570985
Totality Starts at 1:27 / Lasts 2:10

51 Lock 4 Park
1598 Lock 4 Rd, Gallatin, TN 37066
36.33217 -86.47095
Totality Starts at 1:27 / Lasts 2:39

43 Smith Springs Public Use Area
Smith Springs Rd, Nashville, TN 37013
36.09235 -86.60021
Totality Starts at 1:27 / Lasts 2:00

52 Thompson Park
278 Morrison St, Gallatin, TN 37066
36.38558 -86.43797
Totality Starts at 1:27 / Lasts 2:40

44 Long Hunter State Park
2910 Hobson Pike, Hermitage, TN
36.09374 -86.55692
Totality Starts at 1:27 / Lasts 2:04

53 Municipal Park
160 Municipal Park Dr, Gallatin, TN
36.40345 -86.43828
Totality Starts at 1:27 / Lasts 2:40

Plan to be at your eclipse viewing site at least two hours before totality starts.

Scan for Map

54 Gallatin Civic Center
210 Albert Gallatin Ave
Gallatin, TN 37066
36.40596 -86.43787
Totality Starts at 1:27 / Lasts 2:40

55 Triple Creek Park
Gallatin, TN 37066
36.41293 -86.42698
Totality Starts at 1:27 / Lasts 2:40

56 CP HWY 31E
1571 Scottsville Pike Gallatin, TN 37066
36.41147 -86.41168
Totality Starts at 1:27 / Lasts 2:40

57 Bledsoe Creek State Park
400 Zieglers Fort Rd, Gallatin, TN 37066
36.37639 -86.35254
Totality Starts at 1:27 / Lasts 2:40

58 Shady Cove Campground
1115 Shady Cove Rd,
Castalian Springs, TN 37031
36.37874 -86.34119
Totality Starts at 1:27 / Lasts 2:40

59 Cragfont Museum House
Cragfont Rd, Castalian Springs, TN
36.40470 -86.34218
Totality Starts at 1:27 / Lasts 2:39

60 Bledsoe's Fort Park
Castalian Springs, TN 37031
36.39694 -86.32435
Totality Starts at 1:27 / Lasts 2:39

61 CP 2998 TN-10
Castalian Springs, TN 37031
36.34173 -86.25667
Totality Starts at 1:27 / Lasts 2:40

62 Hartsville Access Area
Hartsville, TN 37074
36.37719 -86.17019
Totality Starts at 1:28 / Lasts 2:37

63 Don Fox Community Park
955 W Baddour Pkwy
Lebanon, TN 37087
36.21836 -86.30814
Totality Starts at 1:28 / Lasts 2:38

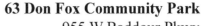

64 Lebanon Airport North End
1086 Leeville Pike
Lebanon, TN 37090
36.19720 -86.31544
Totality Starts at 1:28 / Lasts 2:36

65 Elkins Park - Hobbs Field
713 Elkins Dr, Lebanon, TN 37087
36.19698 -86.30623
Totality Starts at 1:28 / Lasts 2:36

66 Baird Municipal Park
Lebanon, TN 37087
36.19709 -86.28383
Totality Starts at 1:28 / Lasts 2:37

67 Timberline Campground
1204 Murfreesboro Rd
Lebanon, TN 37090
36.17147 -86.30723
Totality Starts at 1:28 / Lasts 2:33

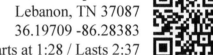

68 Shady Acres RV Park
1639 Murfreesboro Rd
Lebanon, TN 37090
36.15486 -86.30837
Totality Starts at 1:28 / Lasts 2:31

69 Sellars Farm State Park
Lebanon, TN 37090
36.16500 -86.24273
Totality Starts at 1:28 / Lasts 2:36

70 Watertown Community Park
Watertown, TN 37184
36.09695 -86.13834
Totality Starts at 1:28 / Lasts 2:33

71 De Kalb Co Fairgrounds
Fairgrounds Rd, Alexandria, TN 37012
36.07980 -86.03385
Totality Starts at E2:28 / Lasts 2:37

Plan to be at your eclipse viewing site at least two hours before totality starts.

Scan for Map

72 CP US 70N
780 Lebanon Hwy, Lebanon, TN 37087
36.26035 -86.07644
Totality Starts at 1:28 / Lasts 2:40

73 Rome Access Area
Lebanon, TN 37087
36.26163 -86.07188
Totality Starts at 1:28 / Lasts 2:40

74 Carthage City Park
Carthage, TN 37030
36.25817 -85.95791
Totality Starts at 1:28 / Lasts 2:38

75 Cumberland River at Carthage
Upper Ferry Rd, Carthage, TN 37030
36.24937 -85.95284
Totality Starts at 1:28 / Lasts 2:38

76 Defeated Creek Park Campground
140 Marina Ln, Defeated, TN 37030
36.30276 -85.90933
Totality Starts at 1:28 / Lasts 2:34

77 Salt Lick Creek Campground
520 Salt Lick Park Ln
Gainesboro, TN 38562
36.32211 -85.78492
Totality Starts at 1:28 / Lasts 2:30

78 Roaring River Park
Dodson Branch Hwy, Gainesboro, TN
36.37512 -85.63982
Totality Starts at 1:29 / Lasts 2:14

79 Turney Ford Field
Gordonsville, TN 38563
36.17143 -85.93158
Totality Starts at 1:28 / Lasts 2:40

80 CP I-40
Lancaster, TN 38569
36.13863 -85.80855
Totality Starts at 1:29 / Lasts 2:40

Scan for Map

81 Rest Area I-40
Lancaster, TN 38569
36.13982 -85.80674
Totality Starts at 1:29 / Lasts 2:40

82 CP I-40
Silver Point, TN 38582
36.12441 -85.77741
Totality Starts at 1:29 / Lasts 2:40

83 Long Branch Recreation Area
Silver Point, TN 38582
36.10083 -85.83530
Totality Starts at 1:29 / Lasts 2:39

84 Buffalo Valley Recreation Area
Silver Point, TN 38582
36.10148 -85.82891
Totality Starts at 1:29 / Lasts 2:39

85 Center Hill Recreation Area
Liberty, TN 37095
36.09442 -85.83345
Totality Starts at 1:29 / Lasts 2:39

86 Edgar Evins State Park
1630 Edgar Evins State Park Rd
Silver Point, TN 38582
36.08416 -85.82573
Totality Starts at 1:29 / Lasts 2:39

87 Floating Mill Park
Silver Point, TN 38582
36.04532 -85.76549
Totality Starts at 1:29 / Lasts 2:39

88 Hurricane Bridge Rec Area
Silver Point, TN 38582
36.03602 -85.75125
Totality Starts at 1:29 / Lasts 2:39

89 Joe L Evins Park
Smithville, TN 37166
35.95224 -85.82599
Totality Starts at 1:29 / Lasts 2:33

Plan to be at your eclipse viewing site at least two hours before totality starts.

Scan for Map

90 Greenbrook Park
Smithville, TN 37166
35.94952 -85.81853
Totality Starts at 1:29 / Lasts 2:33

91 Ragland Bottom Recreation Area
Sparta, TN 38583
35.97722 -85.72220
Totality Starts at 1:29 / Lasts 2:38

92 Rock Island State Park
Beach Rd, Rock Island, TN 38581
35.81579 -85.64600
Totality Starts at 1:29 / Lasts 2:26

93 CP I-40
Silver Point, TN 38582
36.09573 -85.71463
Totality Starts at 1:29 / Lasts 2:40

94 Love's Travel Stop
110 Fast Lane I-40 Exit 280
Baxter, TN 38544
36.13866 -85.63261
Totality Starts at 1:29 / Lasts 2:36

95 Cummins Falls State Park
1081 Cummins Mill Rd
Cookeville, TN 38501
36.25299 -85.56485
Totality Starts at 1:29 / Lasts 2:26
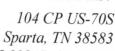

96 Cane Creek Park
201 C C Camp Rd, Cookeville, TN
36.16023 -85.53747
Totality Starts at 1:29 / Lasts 2:33

97 Dogwood Park
E Broad St, Cookeville, TN 38501
36.16444 -85.50341
Totality Starts at 1:29 / Lasts 2:32

98 Parkview Park
Cookeville, TN 38501
36.15240 -85.50983
Totality Starts at 1:29 / Lasts 2:32

Scan for Map

99 Putnam County Fairgrounds
35 Paris St, Cookeville, TN 38501
36.14543 -85.50751
Totality Starts at 1:29 / Lasts 2:33

100 City Lake Natural Area
Bridgeway Dr, Cookeville, TN 38506
36.13333 -85.44726
Totality Starts at 1:29 / Lasts 2:32

101 Burgess Falls State Park
4000 Burgess Falls Dr, Sparta, TN 38583
36.04363 -85.59369
Totality Starts at 1:29 / Lasts 2:40

102 Upper Cumberland Airport
Sparta, TN 38583
36.04924 -85.53599
Totality Starts at 1:29 / Lasts 2:39

103 Golden Mountain Park LLC
6320 Golden Mountain Rd
Sparta, TN 38583
36.02187 -85.48303
Totality Starts at 1:29 / Lasts 2:39

104 CP US-70S
Sparta, TN 38583
35.99847 -85.50249
Totality Starts at 1:29 / Lasts 2:40

105 Whitaker Park
710 E Commercial Ave
Monterey, TN 38574
36.14293 -85.26078
Totality Starts at 1:30 / Lasts 2:22

106 Sparta Rock House
3663 Country Club Rd
Sparta, TN 38583
35.92066 -85.40408
Totality Starts at 1:30 / Lasts 2:40

107 CP Bockman Way
Sparta, TN 38583
35.93793 -85.37103
Totality Starts at 1:30 / Lasts 2:40

Plan to be at your eclipse viewing site at least two hours before totality starts.

Scan for Map

108 Billy Branch Lake
533 Billys Branch Rd, Sparta, TN 38583
35.88446 -85.33074
Totality Starts at 1:30 / Lasts 2:40

109 Virgin Falls State Natural Area
Sparta, TN 38583
35.85420 -85.28204
Totality Starts at 1:30 / Lasts 2:39

110 Fall Creek State Park
10821 Park Rd, Spencer, TN 37367
35.67888 -85.33935
Totality Starts at 1:30 / Lasts 2:27

111 Mountain Glen RV Park
6182 Brockdell Rd, Dunlap, TN 37327
35.55773 -85.38795
Totality Starts at 1:31 / Lasts 2:03

112 Fox Creek Lake
Fox Creek Rd,Crossville, TN 38571
35.99710 -84.97001
Totality Starts at 1:31 / Lasts 2:23

113 Bean Pot Campground
23 Bean Pot Campground Loop
Crossville, TN 38571
35.98004 -84.96546
Totality Starts at 1:31 / Lasts 2:25

114 Centennial Park
Crossville, TN 38555
35.96607 -85.02945
Totality Starts at 1:30 / Lasts 2:30

115 Lake Holiday
Holiday Dr, Crossville, TN 38555
35.95683 -85.05995
Totality Starts at 1:30 / Lasts 2:32

116 Crossville Memorial Airport
Crossville, TN 38555
35.94935 -85.09327
Totality Starts at 1:30 / Lasts 2:33

Scan for Map

117 Cumberland Mountain State Park
24 Office Dr, Crossville, TN 38555
35.90164 -84.99749
Totality Starts at 1:31 / Lasts 2:33

118 Tansi Beach at Lake Tansi
2036 Cravens Dr, Crossville, TN 38572
35.86617 -85.063278
Totality Starts at 1:30 / Lasts 2:36

119 Lake Tansi RV Park
182 Commanche Trail, Crossville, TN
35.86280 -85.07175
Totality Starts at 1:30 / Lasts 2:37

120 Breckenridge Lake RV Park
395 Oak Park Cir, Crossville, TN
35.81058 -85.06291
Totality Starts at 1:31 / Lasts 2:39

121 CP 13047 TN-28 Scenic
Pikeville, TN 37367
35.78105 -85.03244
Totality Starts at 1:31 / Lasts 2:40

122-158 Start Times are E = Eastern Daylight Time

122 Flour Mill Flats Ball Field
609 N Roane St, Harriman, TN 37748
35.93780 -84.55615
Totality Starts at E2:32 / Lasts 2:00

123 Riverfront Park
Emory Dr, Harriman, TN 37748
35.92820 -84.55173
Totality Starts at E2:32 / Lasts 2:02

124 Kingston City Park
333 W Race St, Kingston, TN 37763
35.87908 -84.52665
Totality Starts at E2:32 / Lasts 2:11

125 Kingston Waterfront Park
214 S Kentucky St, Kingston, TN 37763
35.86773 -84.52052
Totality Starts at E2:32 / Lasts 2:12

Plan to be at your eclipse viewing site at least two hours before totality starts.

Scan for Map

126 Southwest Point Park
1218 S Kentucky St
Kingston, TN 37763
35.85844 -84.52660
Totality Starts at E2:32 / Lasts 2:14

135 Engleman Park
Fair St, Sweetwater, TN 37874
35.60100 -84.47570
Totality Starts at E2:32 / Lasts 2:36

127 Caney Creek RV Resort
3615 Roane State Hwy
Harriman, TN 37748
35.86513 -84.59625
Totality Starts at E2:32 / Lasts 2:18

136 Duck Park
Old Hwy 68, Sweetwater, TN 37874
35.59810 -84.46264
Totality Starts at E2:32 / Lasts 2:35

128 CP US-27
23012 Rhea County Hwy
Spring City, TN 3738
35.69860 -84.85558
Totality Starts at E2:31 / Lasts 2:39

137 CP I-75
Niota, TN 37826
35.55777 -84.55543
Totality Starts at E2:32 / Lasts 2:39

129 Spring City Park
Spring City, TN 37381
35.68921 -84.85291
Totality Starts at E2:31 / Lasts 2:39

138 CP 2227 US-11
Niota, TN 37826
35.54145 -84.52059
Totality Starts at E2:32 / Lasts 2:39

130 Rhea Springs Recreation Area
Spring City, TN 37381
35.68256 -84.82784
Totality Starts at E2:31 / Lasts 2:40

139 Athens Regional Park
Regional Park Dr, Athens, TN 37303
35.45703 -84.64118
Totality Starts at E2:32 / Lasts 2:37

131 Cedar Point RV Park & Cabins
5265 Wolf Creek Rd, Spring City, TN
35.65068 -84.84986
Totality Starts at E2:31 / Lasts 2:39

140 McMinn Field
Athens, TN 37303
35.43725 -84.60119
Totality Starts at E2:32 / Lasts 2:37

132 Meigs County Park
Decatur, TN 37322
35.63762 -84.77745
Totality Starts at E2:31 / Lasts 2:39

141 Veterans Memorial Park
Athens, TN 37303
35.43880 -84.58519
Totality Starts at E2:32 / Lasts 2:38

133 Agency Creek Campground
4710 TN-58, Decatur, TN 37322
35.37052 -84.90696
Totality Starts at 1:32 / Lasts 2:09

142 Knox Park
Athens, TN 37303
35.45108 -84.57730
Totality Starts at E2:32 / Lasts 2:38

134 Rock Springs Park
Rock Springs Rd, Lenoir City, TN 37771
35.79650 -84.26598
Totality Starts at E2:32 / Lasts 2:01

143 McMinn County Airport
195 Co Rd 552, Athens, TN 37303
35.39829 -84.56211
Totality Starts at E2:32 / Lasts 2:36

Plan to be at your eclipse viewing site at least two hours before totality starts.

Scan for Map

144 CP 5554 US-411
Madisonville, TN 37354
35.48478 -84.40069
Totality Starts at E2:32 / Lasts 2:39

145 Kefauver Park
Madisonville, TN 37354
35.50047 -84.38750
Totality Starts at E2:32 / Lasts 2:39

146 Monroe County Airport
350 Airport Rd, Sweetwater, TN 37874
35.54621 -84.37918
Totality Starts at E2:32 / Lasts 2:36

147 Vonore Recreation Area
Vonore, TN 37885
35.59061 -84.25240
Totality Starts at E2:33 / Lasts 2:30

148 Fort Loudon State Historic Park
338 Fort Loudon Rd, Vonore, TN 37885
35.58906 -84.21385
Totality Starts at E2:33 / Lasts 2:30

149 Sequoyah Birthplace Memorial
Vonore, TN 37885
35.58062 -84.21827
Totality Starts at E2:33 / Lasts 2:30

150 McGee Carson Peninsula State Historic Park
Vonore, TN 37885
35.58166 -84.19930
Totality Starts at E2:33 / Lasts 2:30

151 Toqua Campground
1315 TN-360, Vonore, TN 37885
35.56641 -84.19274
Totality Starts at E2:33 / Lasts 2:30

152 Toqua Recreation Area
Vonore, TN 37885
35.56071 -84.19538
Totality Starts at E2:33 / Lasts 2:31

Scan for Map

153 Notchy Creek Recreation Area
Scenic River Rd, Madisonville, TN
35.52552 -84.22600
Totality Starts at E2:33 / Lasts 2:33

154 Little Tennessee River at Tallassee
Calderwood Hwy, Tallassee, TN 37878
35.54727 -84.06669
Totality Starts at E2:33 / Lasts 2:26

155 Abrams Falls Trailhead
Townsend, TN 37882
35.59127 -83.85203
Totality Starts at E2:34 / Lasts 2:03

156 Cherohala Skyway Visitor Center
225 Cherohala Skyway
Tellico Plains, TN 37385
35.36694 -84.29686
Totality Starts at E2:33 / Lasts 2:38

157 Bald River Falls
River Rd, Tellico Plains, TN 37385
35.32473 -84.17782
Totality Starts at E2:33 / Lasts 2:38

158 Indian Boundary Lake Recreation Area
Tellico Plains, TN 37385
35.40122 -84.11183
Totality Starts at E2:33 / Lasts 2:37

Plan to be at your eclipse viewing site at least two hours before totality starts.

78 Places to See the Total Eclipse in North Carolina and Georgia

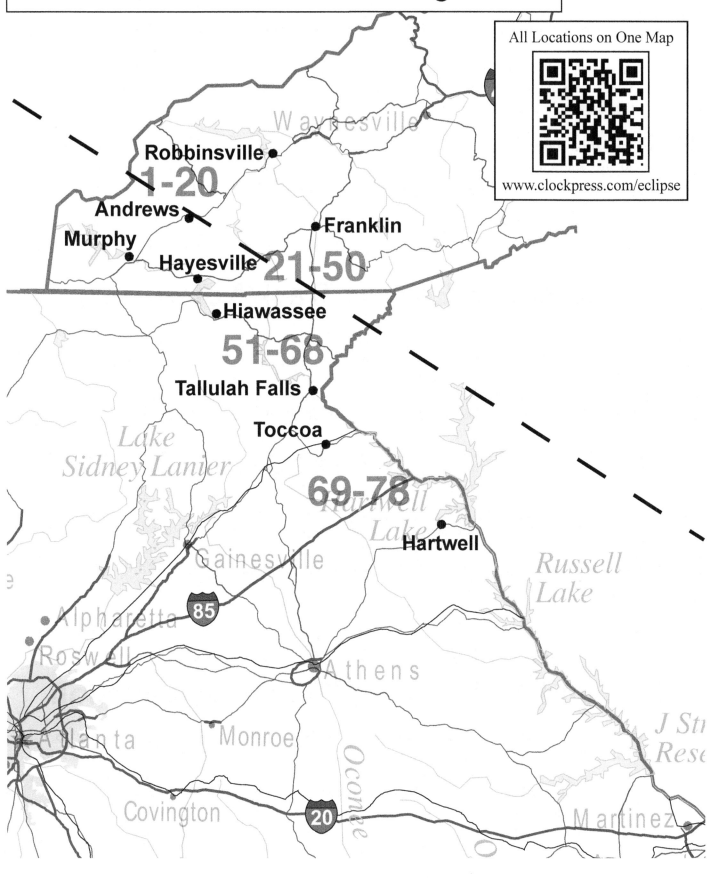

All Locations on One Map

www.clockpress.com/eclipse

Waynesville

Robbinsville

1-20

Andrews

Murphy

Franklin

Hayesville

21-50

Hiawassee

51-68

Tallulah Falls

Toccoa

Lake
Sidney Lanier

69-78

Hartwell Lake

Hartwell

Gainesville

Russell Lake

85

Alpharetta

Roswell

Athens

Atlanta

Monroe

J St
Rese

Covington

20

Martinez

North Carolina Eclipse Track Notes

How many miles long is the eclipse centerline in North Carolina? **42 miles**

What is the average duration of totality along the centerline in North Carolina? **2:39 (m:ss)**
How long does totality last on the centerline at the Tennessee/North Carolina border? **2:39 (m:ss)**
How long does totality last on the centerline at the North Carolina/Georgia border? **2:38 (m:ss)**

When does the partial eclipse start on the centerline at the Tennessee/North Carolina border? **1:05 p.m. (EDT)**
When does totality start on the centerline at the Tennessee/North Carolina border? **2:33 p.m. (EDT)**
When does the partial eclipse end on the centerline at the Tennessee/North Carolina border? **3:59 p.m. (EDT)**

When does the partial eclipse start on the centerline at the North Carolina/Georgia border? **1:06 p.m. (EDT)**
When does totality start on the centerline at the North Carolina/Georgia border? **2:35 p.m. (EDT)**
When does the partial eclipse end on the centerline at the North Carolina/Georgia border? **4:01 p.m. (EDT)**

Check local forecasts and satellite maps for the best weather information for this eclipse.

Georgia Eclipse Track Notes

How many miles long is the eclipse centerline in Georgia? **13 miles**

What is the average duration of totality along the centerline in Georgia? **2:38 (m:ss)**
How long does totality last on the centerline at the North Carolina/Georgia border? **2:38 (m:ss)**
How long does totality last on the centerline at the Georgia/South Carolina border? **2:38 (m:ss)**

When does the partial eclipse start on the centerline at the North Carolina/Georgia border? **1:06 p.m. (EDT)**
When does totality start on the centerline at the North Carolina/Georgia border? **2:35 p.m. (EDT)**
When does the partial eclipse end on the centerline at the North Carolina/Georgia border? **4:01 p.m. (EDT)**

When does the partial eclipse start on the centerline at the Georgia/South Carolina border? **1:07 p.m. (EDT)**
When does totality start on the centerline at the Georgia/South Carolina border? **2:36 p.m. (EDT)**
When does the partial eclipse end on the centerline at the Georgia/South Carolina border? **4:01 p.m. (EDT)**

Check local forecasts and satellite maps for the best weather information for this eclipse.

North Carolina and Georgia
Cities and Highways in Totality by Location

1-20: North of US-74 in North Carolina - **US-129, US-74** Fontana Dam, Robbinsville, Topton, Andrews, Murphy

21-49: US-64 Corridor in North Carolina - **US-23, US-64** Hayesville, Franklin, Otto, Highlands, Lake Toxaway

50-78: Northeastern corner of Georgia - **US-76, US-23, US-123, I-85** - Hiawassee, Clarkesville, Dillard, Mountain City, Lakemont, Tallulah Falls, Toccoa, Lavonia, Hartwell

Featured Eclipse Destinations in North Carolina and Georgia

Gorges State Park in North Carolina has spectacular views, camping, picnic areas and hiking. Over two minutes of totality will be seen from this park on eclipse day. Totality Starts at 2:36 p.m. EDT. See North Carolina #38 for location details.

Website

Photos

Tallulah Gorge State Park in far northeast Georgia is spread over 2,500 acres featuring a suspension bridge, picnic areas, campground, scenic overlooks and much more. Two minutes and twenty three seconds of totality can be seen from this park on eclipse day. Totality Starts at 2:36 p.m. EDT. See Georgia #67 for details.

Website

Photos

North Carolina and Georgia Weather and Eclipse Related Links

Eastern Tennessee Weather
Morristown, TN NWS Office
http://www.srh.noaa.gov/mrx/

Website

Mid-Atlantic
U.S. GOES Satellite - Visible Loop
http://www.ssd.noaa.gov/goes/east/ma/h5-mloop-vis.html

Website

South Carolina Weather
Greenville - Spartanburg, SC
National Weather Service Office
http://www.weather.gov/gsp/

Website

NC / GA Crossing Points Table

Where the Eclipse Centerline Crosses Highways in North Carolina and Georgia				
			All Start Times are Eastern Daylight Time	
Loc#	Hwy	Nearest Mile Marker, Cross Street or Exit, City	TStart	TLasts
18	US-74	Robbinsville Rd, Andrews, NC	2:34	2:39
26	US-64	Black Creek Dr, Franklin, NC	2:35	2:39
43	US-23	Wallalieu Gap Rd, Otto, NC	2:35	2:38
56	GA-246	Kelly's Creek Rd, Dillard, GA	2:35	2:38

North Carolina

Scan for Map   **All Start Times are Eastern Daylight Time**

 1 Fontana Dam at Fontana Lake
Fontana Dam, NC 28733
35.45171 -83.80173
Totality Starts at 2:34 / Lasts 2:22

 2 Cable Cove on Fontana Lake
Cable Cove Rd, Fontana Dam, NC
35.43762 -83.74784
Totality Starts at 2:34 / Lasts 2:21

 3 Huckleberry Knob Summit
Cherohala Skyway Turnout
Cheoah, NC 28771
35.32226 -83.99350
Totality Starts at 2:33 / Lasts 2:38

 4 Cheoah Point Campground
Lake Santeetlah, Robbinsville, NC
35.36988 -83.87158
Totality Starts at 2:34 / Lasts 2:32

5 Deyton Camp
Robbinsville, NC 28771
35.34259 -83.81701
Totality Starts at 2:34 / Lasts 2:32

 6 Simple Life Campground & Cabins
88 Lower Mountain Creek Rd
Robbinsville, NC 28771
35.33572 -83.80505
Totality Starts at 2:34 / Lasts 2:32

 7 Stecoah Valley RV Resort
415 Hyde Town Rd, Robbinsville, NC
35.36966 -83.68391
Totality Starts at 2:34 / Lasts 2:25

 8 Turkey Creek Campground
135 Turkey Creek Rd, Almond, NC
35.37780 -83.57401
Totality Starts at 2:34 / Lasts 2:17

Scan for Map
 9 Fields of the Wood Bible Park
10000 NC-294, Murphy, NC 28906
35.12111 -84.25085
Totality Starts at 2:33 / Lasts 2:19

 10 Hiwassee Dam
State Rd 1314 Murphy, NC 28906
35.15304 -84.17906
Totality Starts at 2:33 / Lasts 2:28

 11 Crawford's Campground
87 Horton Rd, Murphy, N.C., 28906
35.20077 -84.08471
Totality Starts at 2:33 / Lasts 2:37

 12 Persimmon Creek Campground
200 Sunny Point Rd, Murphy, NC 28906
35.03188 -84.20862
Totality Starts at 2:34 / Lasts 2:08

 13 Rivers Edge Mountain RV Resort
1750 Hilltop Rd, Murphy, NC 28906
35.02895 -84.11567
Totality Starts at 2:34 / Lasts 2:14

 14 Cherokee County Recreation Area
699 Conehete St, Murphy, NC 28906
35.09299 -84.02665
Totality Starts at 2:34 / Lasts 2:30

 15 Murphy / Peace Valley KOA
117 Happy Valley Rd, Marble, NC 28905
35.12418 -83.99144
Totality Starts at 2:34 / Lasts 2:35

 16 Western Carolina Regional Airport
5840 Airport Rd, Andrews, NC 28901
35.19666 -83.86278
Totality Starts at 2:34 / Lasts 2:39

 17 Andrews Recreation Park
160 Park St, Andrews, NC 28901
35.20209 -83.83217
Totality Starts at 2:34 / Lasts 2:39

Plan to be at your eclipse viewing site at least two hours before totality starts.

Scan for Map

18 CP US-74
Andrews, NC 28901
35.20683 -83.81684
Totality Starts at 2:34 / Lasts 2:39

27 Standing Indian Campground
2037 West Main St, Franklin, NC 28734
35.07612 -83.52858
Totality Starts at 2:35 / Lasts 2:38

Scan for Map

19 Nelson's Nantahala Hideaway
23797 US-19 Topton, NC 28781
35.24494 -83.70286
Totality Starts at 2:34 / Lasts 2:35

28 Hurricane Creek
Primative Campground
Smithbridge, NC
35.05537 -83.50985
Totality Starts at 2:35 / Lasts 2:39

20 Ferebee Memorial Picnic Area
W Hwy 19, Bryson City, NC 28713
35.30319 -83.65356
Totality Starts at 2:34 / Lasts 2:30

29 Albert Mountain Fire Tower
Otto, NC 28763
35.05262 -83.47758
Totality Starts at 2:35 / Lasts 2:39

21 Clay County Recreation Center
Ball Park Dr, Hayesville, NC 28904
35.04617 -83.81087
Totality Starts at 2:34 / Lasts 2:36

30 Appalachian Trail Trail Head
Hwy 64 Winding Stair Gap
Franklin, NC 28734
35.11402 -83.55094
Totality Starts at 2:35 / Lasts 2:37

22 Clay County Park Campground
47 Clay Recreation Park Rd
Hayesville, NC 28904
35.00590 -83.79799
Totality Starts at 2:34 / Lasts 2:32

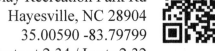

31 Cartoogechaye Creek Campgrnd
91 No Name Rd, Franklin, NC 28734
35.13698 -83.49064
Totality Starts at 2:35 / Lasts 2:35

23 Chatuge Dam
1407 Chatuge Dam Rd
Hayesville, NC 28904
35.02045 -83.78547
Totality Starts at 2:34 / Lasts 2:34

32 Pines RV Park and Cabins
4724 Murphy Rd, Franklin, NC 28734
35.14869 -83.46712
Totality Starts at 2:35 / Lasts 2:33

24 Jackrabbit Campground
1536 Jack Rabbit Rd,
Hayesville, NC 28904
35.01129 -83.77277
Totality Starts at 2:34 / Lasts 2:34

33 Macon County Airport
Franklin, NC 28734
35.22003 -83.42752
Totality Starts at 2:35 / Lasts 2:28

25 Ho-Hum Campground
47 Ho Hum Loop, Hayesville, NC 28904
34.99643 -83.75668
Totality Starts at 2:34 / Lasts 2:33

34 Mi Mountain Campground
151 Mi Mountain Rd
Franklin, NC 28734
35.22098 -83.35279
Totality Starts at 2:35 / Lasts 2:24

26 CP US-64 / Murphy Rd
Franklin, NC 28734
35.08858 -83.57079
Totality Starts at 2:35 / Lasts 2:39

35 Ralph J Andrews Recreation Park
814 Ralph Andrews Park Rd
Glenville, NC 28736
35.19186 -83.15446
Totality Starts at 2:35 / Lasts 2:15

Plan to be at your eclipse viewing site at least two hours before totality starts.

Scan for Map

Scan for Map

36 Panther Ridge RV Park Camp
13 Little Panthertail Rd
Lake Toxaway, NC 28747
35.15928 -82.94730
Totality Starts at 2:36 / Lasts 2:01

45 Osage Lake
190 Lakeside Rd, Franklin, NC 28734
35.00673 -83.29203
Totality Starts at 2:35 / Lasts 2:36

37 River Bend RV Park
1476 Blue Ridge Rd
Lake Toxaway, NC 28747
35.14556 -82.92832
Totality Starts at 2:36 / Lasts 2:02

46 Vanhook Glade Campground
14014 Highlands Rd
Highlands, NC 28741
35.07731 -83.24840
Totality Starts at 2:35 / Lasts 2:31

38 Gorges State Park Visitor Center
976 Grassy Ridge Rd
Sapphire, NC 28774
35.09663 -82.95039
Totality Starts at 2:36 / Lasts 2:14

47 Cliffside Lake Recreation Area
13908 Highlands Rd
Highlands, NC 28741
35.07893 -83.23687
Totality Starts at 2:35 / Lasts 2:31

39 Macon County Recreation Park
1288 Georgia Rd, Franklin, NC 28734
35.15757 -83.39065
Totality Starts at 2:35 / Lasts 2:31

48 Bridal Veil Falls
Hwy 64, Highlands NC 28741
35.07219 -83.22914
Totality Starts at 2:35 / Lasts 2:31

40 Old Millsite Campground
80-33 Nickajack Rd Franklin, NC 28734
35.14411 -83.30823
Totality Starts at 2:35 / Lasts 2:30

49 Whiteside Mountain Trail
170 Deville Dr, Highlands, NC 28741
35.08052 -83.14407
Totality Starts at 2:36 / Lasts 2:28

41 Franklin RV Park & Campground
230 Addington Bridge Rd,
Franklin, NC 28734
35.12360 -83.39590
Totality Starts at 2:35 / Lasts 2:33

Georgia

50 Poteete Creek Camp Ground
1040 Poteete Creek Rd
Blairsville, GA 30512
34.94847 -84.09488
Totality Starts at 2:34 / Lasts 2:03

42 Smoky Mountain Host
4437 Georgia Rd Franklin, NC 28734
35.11264 -83.39104
Totality Starts at 2:35 / Lasts 2:33

*43 CP HWY 441 Georgia Rd,
Otto, NC 28763
34.99729 -83.38181
Totality Starts at 2:35 / Lasts 2:38*

51 Chatuge Woods Campground
2307 Red Banks Dr
Young Harris, GA 30582
34.97981 -83.81026
Totality Starts at 2:34 / Lasts 2:29

44 Highlands Aerial Park
9625 Dillard Rd,
Scaly Mountain, NC 28775
35.00553 -83.33174
Totality Starts at 2:35 / Lasts 2:37

52 Towns County Park
Hiawassee, GA 30546
34.96817 -83.77078
Totality Starts at 2:34 / Lasts 2:29

Plan to be at your eclipse viewing site at least two hours before totality starts.

Scan for Map

53 Lake Chatuge Recreation Area
Hiawassee, GA 30546
34.95356 -83.78145
Totality Starts at 2:34 / Lasts 2:27

54 Riverbend Campground
2626 Streak Hill Rd
Hiawassee, GA 30546
34.91083 -83.71173
Totality Starts at 2:35 / Lasts 2:26

55 River Vista Mountain Village
20 River Vista Dr, Dillard, GA 30537
34.98631 -83.36840
Totality Starts at 2:35 / Lasts 2:38

56 CP GA-246 Larry McClure Hwy
Dillard, GA 30537
34.98824 -83.36314
Totality Starts at 2:35 / Lasts 2:38

57 Permisson Valley Campground
Clayton, GA 30525
34.92628 -83.50617
Totality Starts at 2:35 / Lasts 2:37

58 Jones Bridge Park
Lookout Mountain Scenic Hwy
Clayton, GA 30525
34.86867 -83.54103
Totality Starts at 2:35 / Lasts 2:31

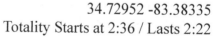

59 Moccasin Creek State Park
3655 GA-197, Clarkesville, GA 30523
34.84469 -83.58726
Totality Starts at 2:35 / Lasts 2:25

60 Sugar Mill Creek RV Resort
4960 Laurel Lodge Rd
Clarkesville, GA 30523
34.79944 -83.56991
Totality Starts at 2:35 / Lasts 2:19

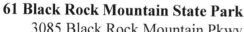

61 Black Rock Mountain State Park
3085 Black Rock Mountain Pkwy
Mountain City, GA 30562
34.90601 -83.40667
Totality Starts at 2:35 / Lasts 2:37

Scan for Map

62 Cross Creek Camp Ground
618 File St, Mountain City, GA 30562
34.90950 -83.38752
Totality Starts at 2:35 / Lasts 2:37

63 Rabun County Park
400 Rabun County Park Rd
Clayton, GA 30525
34.82212 -83.42209
Totality Starts at 2:35 / Lasts 2:32

64 Lake Rabun Beach Rec Area
5315 Lake Rabun Rd
Lakemont, GA 30552
34.75633 -83.48491
Totality Starts at 2:35 / Lasts 2:19

65 Persimmon Point on Lake Rabun
49 Piney Point, Lakemont, GA 30552
34.77205 -83.43021
Totality Starts at 2:35 / Lasts 2:25

66 Terrora Park Campground
300 Jane Hurt Yarn Rd
Tallulah Falls, GA 30573
34.73989 -83.39787
Totality Starts at 2:36 / Lasts 2:23

67 Tallulah Gorge State Park
338 Jane Hurt Yarn Rd
Tallulah Falls, GA 30573
34.74012 -83.38982
Totality Starts at 2:36 / Lasts 2:23

68 Tallulah Point Overlook
940 Tallulah Gorge Scenic Loop
Tallulah Falls, GA 30573
34.72952 -83.38335
Totality Starts at 2:36 / Lasts 2:22

69 Alewine Park
Alewine Dr, Toccoa, GA 30577
34.58413 -83.34752
Totality Starts at 2:36 / Lasts 2:01

70 Henderson Falls Park
672 Henderson Falls Rd
Toccoa, GA 30577
34.59057 -83.33268
Totality Starts at 2:36 / Lasts 2:04

Plan to be at your eclipse viewing site at least two hours before totality starts.

Scan for Map

71 CP Warwoman Rd
Clayton, GA 30525
34.92210 -83.22677
Totality Starts at 2:36 / Lasts 2:38

72 Poplar Springs Recreation Area
Poplar Springs Recreation Area Rd
Lavonia, GA 30553
34.52136 -83.08996
Totality Starts at 2:37 / Lasts 2:12

73 Tugaloo State Park
1763 Tugaloo State Park Rd
Lavonia, GA 30553
34.49293 -83.06262
Totality Starts at 2:37 / Lasts 2:09

74 Paynes Creek Campground
518 Ramp Rd, Hartwell, GA 30643
34.47578 -82.97681
Totality Starts at 2:37 / Lasts 2:13

75 Hart State Park
232 Hart State Park Rd
Hartwell, GA 30643
34.38286 -82.90941
Totality Starts at 2:37 / Lasts 2:03

76 Long Point Recreation Area
Old 29 Hwy, Hartwell, GA 30643
34.38201 -82.85326
Totality Starts at 2:37 / Lasts 2:08

77 Elrod Ferry Recreation Area
Elrod Ferry Rd, Hartwell, GA 30643
34.36181 -82.85468
Totality Starts at 2:37 / Lasts 2:04

78 Watsadler Campground
286 Watsadler Rd, Hartwell, GA 30643
34.34189 -82.84227
Totality Starts at 2:38 / Lasts 2:02

Plan to be at your eclipse viewing site at least two hours before totality starts.

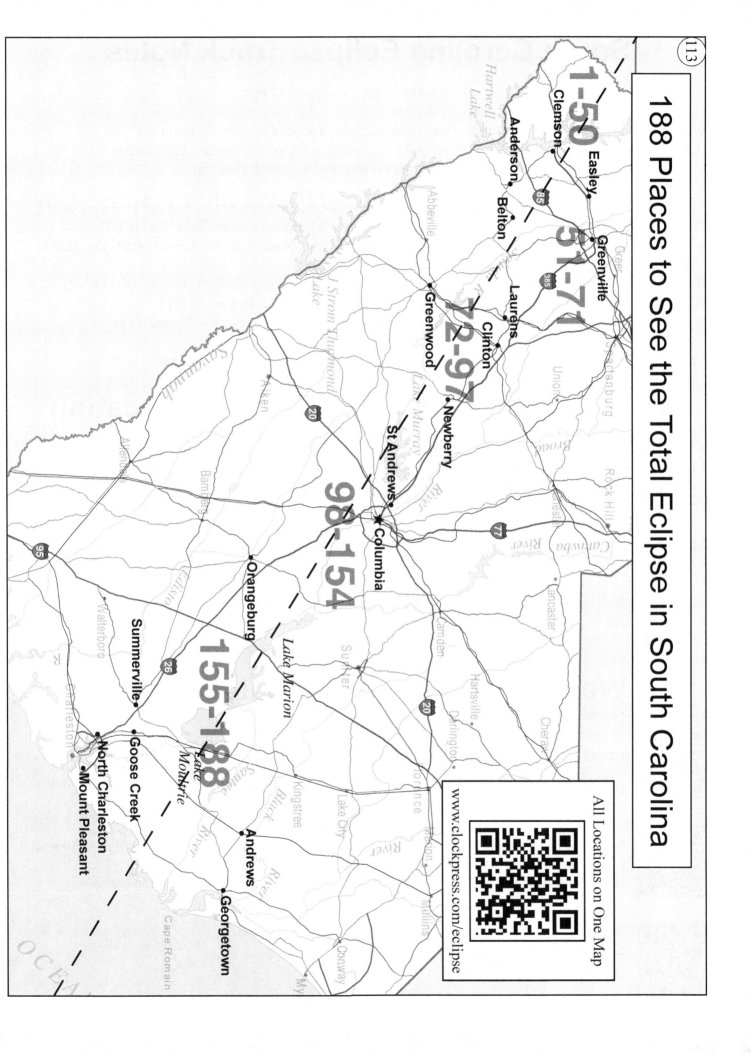

188 Places to See the Total Eclipse in South Carolina

All Locations on One Map

www.clockpress.com/eclipse

South Carolina Eclipse Track Notes

How many miles long is the eclipse centerline in South Carolina? **247 miles**

What is the average duration of totality along the centerline in South Carolina? **2:36 (m:ss)**
How long does totality last on the centerline at the Georgia/South Carolina border? **2:38 (m:ss)**
How long does totality last on the centerline at the South Carolina coast? **2:34 (m:ss)**

When does the partial eclipse start on the centerline at the Georgia/South Carolina border? **1:07 p.m. (EDT)**
When does totality start on the centerline at the Georgia/South Carolina border? **2:36 p.m. (EDT)**
When does the partial eclipse end on the centerline at the Georgia/South Carolina border? **4:01 p.m. (EDT)**

When does the partial eclipse start on the centerline at the South Carolina coast? **1:17 p.m. (EDT)**
When does totality start on the centerline at the South Carolina coast? **2:46 p.m. (EDT)**
When does the partial eclipse end on the centerline at the South Carolina coast? **4:10 p.m. (EDT)**

Check local forecasts and satellite maps for the best weather information for this eclipse.

Cities and Highways in Totality by Location

1-50: North and West of Greenville - **US-178, US-123, I-85, SC-93, US-76, US-29** - Seneca, Clemson, Fair Play, Anderson, Pendleton, Pickens, Easley, Belton, Piedmont

51-71: Greenville - **US-276, US-29, US-123, US-25, I-85, I-185, I-385**

72-108: South of Greenville, North of Columbia - **I-385, I-26, US-76** - Fountain Inn, Honea Path, Ware Shoals, Laurens, Newberry, Greenwood, Clinton, Irmo

109-133: Columbia - **I-26, I-20, I-126, I-77, US-21, US-378, US-321, US-1**

134-188: South of Columbia to the Coast - **US-378, I-26, US-601, US-301, US-178, US-176, I-95, US-52, I-526, US-17, US-521** - Hopkins, Gaston, Pelion, Orangeburg, Santee, Pinopolis, Moncks Corner, St. Stephen, Andrews, Summerville, North Charleston, Mt. Pleasant, McClellanville, Georgetown

Weather and Eclipse Related Links

Website

South Carolina Weather
Greenville - Spartanburg, SC
National Weather Service Office
http://www.weather.gov/gsp/

Mid-Atlantic
U.S. GOES Satellite - Visible Loop
http://www.ssd.noaa.gov/goes/east/ma/h5-mloop-vis.html

Middle South Carolina Weather
Columbia, SC NWS Office
http://www.weather.gov/cae/

Columbia, SC Convention and Visitors Bureau
Solar Eclipse Site
http://www.columbiacvb.com/events/solar-eclipse/

Coastal South Carolina Weather
Charleston, SC NWS Office
http://www.weather.gov/chs/

Featured Eclipse Destinations

Oconee State Park has a large campground with over 150 sites. The park sits on over 1,000 acres and includes a picnic area, playground and more. Eclipse watchers can see two minutes and thirty eight seconds of totality from here on eclipse day. Totality Starts at 2:36 p.m. EDT. See South Carolina #5 for location details.

Website

Photos

Dreher Island State Park on Lake Murray is a full featured park with a dozen picnic shelters, over 100 campsites, boating, fishing and several lakeside villas. The total eclipse lasts over two and a half minutes at this location. Totality Starts at 2:40 p.m. EDT. See South Carolina #96 for location details.

Website

Photos

South Carolina Botanical Garden offers nearly 300 acres of walking trails, streams and gardens. Visitors can experience over two and a half minutes of totality during the eclipse. Totality Starts at 2:37 p.m. EDT. See South Carolina #26 for details.

Website

Photos

Santee State Park is situated on Lake Marion, the largest lake in South Carolina. Picnic shelters and hiking trails are available. The eclipse brings two minutes and thirty six seconds of totality to this large park. Totality Starts at 2:43 p.m. EDT. See South Carolina #154 for location details.

Website

Photos

SC Crossing Points Table

		Where the Eclipse Centerline Crosses Highways in South Carolina		
			All Start Times are Eastern Daylight Time	
Loc#	**Hwy**	**Nearest Mile Marker, Cross Street or Exit, City**	**TStart**	**TLasts**
6	SC-107	Courtenay King Dr, Mountain Rest, SC	2:36	2:38
29	SC-93	Pecan St, Central, SC	2:37	2:38
31	US-123	MM 5, State Rd S-39-18, Central, SC	2:37	2:38
33	US-178	Allgood Rd, Pendleton, SC	2:37	2:38
35	I-85	MM 26, Scotts Bridge Rd, Anderson, SC	2:37	2:38
44	US-29	Cheddar Rd, Belton, SC	2:37	2:38
76	US-76	Horseshoe Rd, Honea Path, SC	2:38	2:37
78	US-25	Harmony Rd, Ware Shoals, SC	2:38	2:37
82	US-221	Todd Quarter Rd, Waterloo, SC	2:39	2:37
99	US-378	Keisler Rd, Gilbert, SC	2:41	2:37
100	US-1	Sandy Bank Dr, Lexington, SC	2:41	2:36
106	I-20	Exit 55, Lexington, SC	2:41	2:36
138	US-321	Heather Ridge Dr, Gaston, SC	2:41	2:36
139	I-26	MM 124, Exit 125, Gaston, SC	2:42	2:36
142	US-176	State Rd S-9-212, Swansea, SC	2:42	2:36
146	US-601	Hemlock Rd, St Matthews, SC	2:42	2:36
155	I-95	Exit 102, Lake Marion, Summerton, SC	2:43	2:35
162	SC-45	Lizzie Ave, Cross, SC	2:44	2:35
167	US-52	Fairsprings Rd, Bonneau, SC	2:45	2:35
186	US-17	Doe Hall Plantation Rd, McClellanville, SC	2:46	2:34

NW South Carolina

All Start Times are Eastern Daylight Time

1 Table Rock State Park
Pickens, SC 29671
35.03187 -82.69946
Totality Starts at 2:37 / Lasts 2:03

2 Lake Oolenoy
158 E Ellison Ln, South Carolina 29671
35.01947 -82.69304
Totality Starts at 2:37 / Lasts 2:05

3 Devils Fork State Park
161 Holcombe Cir, Salem, SC 29676
34.95315 -82.94610
Totality Starts at 2:36 / Lasts 2:30

4 Keowee Toxaway State Park
108 Residence Dr, Sunset, SC 29685
34.93086 -82.88655
Totality Starts at 2:36 / Lasts 2:30

5 Oconee State Park
Mountain Rest, SC 29664
34.86698 -83.10387
Totality Starts at 2:36 / Lasts 2:38

6 CP 550 SC-107
Mountain Rest, SC 29664
34.86547 -83.11038
Totality Starts at 2:36 / Lasts 2:38

7 Pickens Jaycee Park
149 N Homestead Rd
Pickens, SC 29671
34.89214 -82.71343
Totality Starts at 2:37 / Lasts 2:25

8 High Falls County Park
671 High Falls Rd, Seneca, SC 29672
34.79640 -82.92877
Totality Starts at 2:36 / Lasts 2:38

9 Chau Ram County Park
1220 Chau Ram Park Rd
Westminster, SC 29693
34.68412 -83.14367
Totality Starts at 2:36 / Lasts 2:31

10 Anderson Park
345 State Rd S-37-106
Westminster, SC 29693
34.66631 -83.09610
Totality Starts at 2:36 / Lasts 2:32

11 South Cove Park
1099 S Cove Rd, Seneca, SC 29672
34.71311 -82.96389
Totality Starts at 2:36 / Lasts 2:37

12 Seneca Recreation
98 W South 4th St, Seneca, SC 29678
34.67860 -82.96515
Totality Starts at 2:36 / Lasts 2:37

13 Lake Hartwell State Park
19138-A South Carolina 11
Fair Play, SC 29643
34.49493 -83.03176
Totality Starts at 2:37 / Lasts 2:12

14 Carolina Landing RV Resort
120 Carolina Landing Dr
Fair Play, SC 29643
34.49740 -82.97808
Totality Starts at 2:37 / Lasts 2:17

15 Singing Pines Recreation Area
6600 US-29, Starr, SC 29684
34.37582 -82.81719
Totality Starts at 2:37 / Lasts 2:10

16 Crescent Group Campground
Campers Way, Starr, SC 29684
34.38029 -82.81750
Totality Starts at 2:37 / Lasts 2:10

17 Sadlers Creek State Park
940 Sadlers Creek Rd
Anderson, SC 29626
34.42503 -82.82506
Totality Starts at 2:37 / Lasts 2:17

Plan to be at your eclipse viewing site at least two hours before totality starts.

Scan for Map

Scan for Map

18 Springfield Campground
Springfield Rd, Anderson, SC 29626
34.44240 -82.82398
Totality Starts at 2:37 / Lasts 2:20

27 Armory Ball Field
141 Pendleton Rd, Clemson, SC 29631
34.67450 -82.81375
Totality Starts at 2:37 / Lasts 2:38

19 River Forks Recreation Area
710 River Forks Rd, Anderson, SC 29626
34.47309 -82.81376
Totality Starts at 2:37 / Lasts 2:25

28 Clemson Park and Community Garden
114 Clemson Park Rd, Clemson, SC
34.68393 -82.80569
Totality Starts at 2:37 / Lasts 2:38

20 Asbury Campground
Asbury Park Rd, Anderson, SC 29625
34.53633 -82.78484
Totality Starts at 2:37 / Lasts 2:34

29 CP Old Greenville Hwy 93
1500 W Main St, Central, SC 29630
34.70991 -82.79209
Totality Starts at 2:37 / Lasts 2:38

21 Coneross Park / Campground
Coneross Park Rd, Townville, SC 29689
34.59054 -82.89770
Totality Starts at 2:37 / Lasts 2:34

30 Southern Wesleyan University
907 Wesleyan Dr, Central, SC 29630
34.72779 -82.76177
Totality Starts at 2:37 / Lasts 2:37

22 Oconee Point Campground
200 Oconee Point Rd, Seneca, SC 29678
34.59689 -82.86646
Totality Starts at 2:37 / Lasts 2:36

31 CP Calhoun Memorial Hwy US-123
Central, SC 29630
34.69895 -82.76978
Totality Starts at 2:37 / Lasts 2:38

23 Oconee County Airport
365 Airport Rd, Seneca, SC 29678
34.67156 -82.88500
Totality Starts at 2:37 / Lasts 2:37

32 Veterans Park
Pendleton, SC 29670
34.64225 -82.77246
Totality Starts at 2:37 / Lasts 2:37

24 Y Beach
276 YMCA Cir, Seneca, SC 29678
34.68181 -82.86197
Totality Starts at 2:37 / Lasts 2:37

33 CP US-178 / 6740 Liberty Hwy
Pendleton, SC 29670
34.66216 -82.69489
Totality Starts at 2:37 / Lasts 2:38

25 Presidents Park
Clemson, SC 29634
34.67916 -82.83201
Totality Starts at 2:37 / Lasts 2:37

34 City of Anderson Recreation Area
Anderson, SC 29621
34.57155 -82.68434
Totality Starts at 2:37 / Lasts 2:37

26 South Carolina Botanical Garden
150 Discovery Ln, Clemson, SC 29631
34.66967 -82.82673
Totality Starts at 2:37 / Lasts 2:37

35 CP I-85 at Scotts Bridge Rd
Anderson, SC 29621
34.62297 -82.61520
Totality Starts at 2:37 / Lasts 2:38

Plan to be at your eclipse viewing site at least two hours before totality starts.

Scan for Map

36 Whitehall Park
Anderson, SC 29625
34.53417 -82.67754
Totality Starts at 2:37 / Lasts 2:36

37 Anderson Regional Airport
5805 Airport Rd, Anderson, SC 29626
34.49543 -82.71031
Totality Starts at 2:37 / Lasts 2:33

38 Equinox Park
Standridge Rd, Anderson, SC 29625
34.51354 -82.67525
Totality Starts at 2:37 / Lasts 2:36

39 Southwood Park
Anderson, SC 29624
34.49131 -82.66495
Totality Starts at 2:37 / Lasts 2:35

**40 American Legion
Memorial Athletic Field**
Anderson, SC 29624
34.49487 -82.64447
Totality Starts at 2:37 / Lasts 2:36

41 Jefferson Avenue Park
898 Quarry StAnderson, SC 29624
34.50253 -82.63902
Totality Starts at 2:37 / Lasts 2:36

42 McFall's Landing
1625 Broadway Lake Rd
Anderson, SC 29621
34.45271 -82.59547
Totality Starts at 2:38 / Lasts 2:35

43 Allen Park
Lakeside Dr, Anderson, SC 29621
34.45931 -82.58989
Totality Starts at 2:38 / Lasts 2:36

*44 CP 4668 US-29
Belton, SC 29627
34.59004 -82.54838
Totality Starts at 2:37 / Lasts 2:38*

Scan for Map

45 Leda Poore Park
Belton, SC 29627
34.54229 -82.49294
Totality Starts at 2:38 / Lasts 2:38

46 Williamston Park Spring
Center St, Williamston, SC 29697
34.61711 -82.47928
Totality Starts at 2:38 / Lasts 2:35

47 Pilot Travel Center
110 1147, Piedmont, SC 29673
34.71130 -82.49752
Totality Starts at 2:37 / Lasts 2:31

48 Woodson Memorial Park
520 Mills Ave, Liberty, SC 29657
34.78391 -82.69579
Totality Starts at 2:37 / Lasts 2:32

49 Hagood Park
199 Park Dr, Easley, SC 29640
34.82798 -82.61616
Totality Starts at 2:37 / Lasts 2:27

50 Kings Park
Oak Cir, Easley, SC 29640
34.83667 -82.59597
Totality Starts at 2:37 / Lasts 2:25

51 Northwest Park
8109 White Horse Rd
Greenville, SC 29617
34.90291 -82.47308
Totality Starts at 2:37 / Lasts 2:07

52 Piney Mountain Park
501 Worley Rd, Greenville, SC 29609
34.88848 -82.39006
Totality Starts at 2:38 / Lasts 2:00

53 Holmes Park
112 Holmes Dr, Greenville, SC 29609
34.87989 -82.36836
Totality Starts at 2:38 / Lasts 2:00

Plan to be at your eclipse viewing site at least two hours before totality starts.

54 Bob Jones University
1700 Wade Hampton Blvd
Greenville, SC 29614
34.87564 -82.36466
Totality Starts at 2:38 / Lasts 2:00

Scan for Map
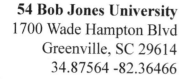
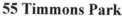

55 Timmons Park
121 Oxford St, Greenville, SC 29607
34.86597 -82.37550
Totality Starts at 2:38 / Lasts 2:04

56 McPherson Park
120 E Park Ave, Greenville, SC 29601
34.85737 -82.39396
Totality Starts at 2:38 / Lasts 2:09

57 Westside Park
2700 W Blue Ridge Dr
Greenville, SC 29611
34.86002 -82.44291
Totality Starts at 2:37 / Lasts 2:12

58 Shoeless Joe Jackson Memorial Park
406 West Ave, Greenville, SC 29611
34.84347 -82.43424
Totality Starts at 2:37 / Lasts 2:14

59 Mayberry Park
70 Mayberry St, Greenville, SC 29601
34.85182 -82.41429
Totality Starts at 2:37 / Lasts 2:11

60 Falls Park on The Reedy
601 S Main St, Greenville, SC 29601
34.84533 -82.40085
Totality Starts at 2:38 / Lasts 2:11

61 Cleveland Park
Cleveland Park Dr, Greenville, SC 29601
34.84459 -82.38711
Totality Starts at 2:38 / Lasts 2:10

62 Greenville Downtown Airport
100 Tower Dr, Greenville, SC 29607
34.85000 -82.34837
Totality Starts at 2:38 / Lasts 2:05

63 Haywood Mall
700 Haywood Rd, Greenville, SC 29607
34.85016 -82.33548
Totality Starts at 2:38 / Lasts 2:03

Scan for Map

64 Legacy Park
336 Rocky Slope Rd
Greenville, SC 29607
34.82980 -82.33101
Totality Starts at 2:38 / Lasts 2:08

65 Dolly Cooper Park
170 Spearman Cir, Greenville, SC 29611
34.80314 -82.47108
Totality Starts at 2:37 / Lasts 2:21

66 7th Inning Splash Park
1500 Piedmont Hwy
Piedmont, SC 29673
34.77437 -82.42960
Totality Starts at 2:37 / Lasts 2:22

67 Donaldson Center Airport
2 Exchange St, Greenville, SC 29605
34.75878 -82.37584
Totality Starts at 2:38 / Lasts 2:21

68 Conestee Park
840 S-23-107, Greenville, SC 29607
34.77931 -82.35219
Totality Starts at 2:38 / Lasts 2:17

69 Sunset Park
211 Fowler Circle, Greenville, SC 29607
34.77216 -82.32473
Totality Starts at 2:38 / Lasts 2:16

70 Springfield Park
Hyde Cir, Mauldin, SC 29662
34.78521 -82.30326
Totality Starts at 2:38 / Lasts 2:13

71 Southside Park
417 Baldwin Rd, Greenville, SC 29607
34.74399 -82.30301
Totality Starts at 2:38 / Lasts 2:18

Plan to be at your eclipse viewing site at least two hours before totality starts.

Scan for Map   Scan for Map

72 Woodside Park
Woodside Ave, Fountain Inn, SC 29644
34.68817 -82.20108
Totality Starts at 2:38 / Lasts 2:19

81 West Cambridge Park
451 Grove St, Greenwood, SC 29646
34.19464 -82.16904
Totality Starts at 2:39 / Lasts 2:29

73 Cedar Falls Park
201 Cedar Falls Rd
Fountain Inn, SC 29644
34.61612 -82.30156
Totality Starts at 2:38 / Lasts 2:31

82 CP 11978 US-221
Waterloo, SC 29384
34.34846 -82.06096
Totality Starts at 2:39 / Lasts 2:37

74 Loretta C. Wood Park
10270 Augusta Rd, Pelzer, SC 29669
34.59431 -82.34974
Totality Starts at 2:38 / Lasts 2:33

83 Lake Greenwood State Park
302 State Park Rd, Ninety Six, SC 29666
34.19501 -81.94934
Totality Starts at 2:39 / Lasts 2:36

75 Honea Path Recreation
Arkansas St, Honea Path, SC 29654
34.44879 -82.38675
Totality Starts at 2:38 / Lasts 2:37

84 Little River Park
E Main St, Laurens, SC 29360
34.49942 -82.00862
Totality Starts at 2:39 / Lasts 2:29

76 CP US-76 / 458 Princeton Hwy
Honea Path, SC 29654
34.48783 -82.34158
Totality Starts at 2:38 / Lasts 2:37

85 Laurens County Airport
Laurens, SC 29360
34.50779 -81.94831
Totality Starts at 2:39 / Lasts 2:25

77 Ware Shoals Dragway
16924 US-25, Ware Shoals, SC 29692
34.45629 -82.25240
Totality Starts at 2:38 / Lasts 2:37

86 Musgrove Mill State Historic Site
398 State Park Rd, Clinton, SC 29325
34.59157 -81.85250
Totality Starts at 2:39 / Lasts 2:07

78 CP 17981 US-25
Ware Shoals, SC 29692
34.44326 -82.25163
Totality Starts at 2:38 / Lasts 2:37

87 New Bailey Stadium
123-197 State Rd S-30-71
Clinton, SC 29325
34.46179 -81.86337
Totality Starts at 2:39 / Lasts 2:26

79 Ware Shoals City Park
Mill St, Ware Shoals, SC 29692
34.39984 -82.24440
Totality Starts at 2:38 / Lasts 2:37

88 Rest Area I-26
Kinards, SC 29355
34.41427 -81.70563
Totality Starts at 2:39 / Lasts 2:22

80 Young Park
104 State Rd S-24-38
Ware Shoals, SC 29692
34.39382 -82.25305
Totality Starts at 2:38 / Lasts 2:37

89 Heritage Park
Park St, Whitmire, SC 29178
34.50126 -81.61549
Totality Starts at 2:40 / Lasts 2:00

Plan to be at your eclipse viewing site at least two hours before totality starts.

Central South Carolina

90 Newberry College

Scan for Map

2100 College St, Newberry, SC 29108
34.28547 -81.62181
Totality Starts at 2:40 / Lasts 2:29

91 Love's Travel Stop

36 Dusty Rd, Newberry, SC 29108
34.29683 -81.54354
Totality Starts at 2:40 / Lasts 2:26

92 Lake Monticello Park

Baltic Cir, Jenkinsville, SC 29065
34.32387 -81.28691
Totality Starts at 2:41 / Lasts 2:03

93 Saluda River Resort

1283 Saluda River Rd
Silverstreet, SC 29145
34.17563 -81.69764
Totality Starts at 2:40 / Lasts 2:37

94 W.L. Mills MD Fitness Park

Prosperity, SC 29127
34.21468 -81.53761
Totality Starts at 2:40 / Lasts 2:31

95 Reunion Park
State Rd S-36-458
Little Mountain, SC 29075
34.19318 -81.41458
Totality Starts at 2:40 / Lasts 2:29

96 Dreher Island State Park
3881 State Park Rd, Prosperity, SC 29127
34.08533 -81.40842
Totality Starts at 2:40 / Lasts 2:34

97 Crooked Creek Park

1098 Old Lexington Hwy
Chapin, SC 29036
34.14318 -81.34204
Totality Starts at 2:41 / Lasts 2:29

Scan for Map

98 Woodsmoke Family Campground

11302 Broad River Rd
Irmo, SC 29063
34.15761 -81.25107
Totality Starts at 2:41 / Lasts 2:25

99 CP 2165 US Route 378

Gilbert, SC 29054
34.01169 -81.38948
Totality Starts at 2:41 / Lasts 2:37

100 CP Hwy 1 / 2738 Augusta Rd

Lexington, SC 29072
33.97911 -81.32492
Totality Starts at 2:41 / Lasts 2:36

101 Watson Park

Watson Park Rd, Johnston, SC 29832
33.82863 -81.81004
Totality Starts at 2:40 / Lasts 2:01

102 Monetta Drive In Theatre

5822 Columbia Hwy N
Monetta, SC 29105
33.85249 -81.59892
Totality Starts at 2:41 / Lasts 2:22

103 Gravatt Camp and Conference Center

Camp Gravatt Rd, Aiken, SC 29805
33.73837 -81.58515
Totality Starts at 2:41 / Lasts 2:05

104 Cedar Pond Campground

4721 Fairview Rd, Leesville, SC 29070
33.80019 -81.43304
Totality Starts at 2:41 / Lasts 2:26

105 Love's Travel Stop

I-20 Exit 51, 340 Longs Pond Rd,
Lexington, SC 29073
33.92436 -81.29160
Totality Starts at 2:41 / Lasts 2:36

106 CP I-20 past Two Notch Rd

Lexington, SC 29073
33.94490 -81.25738
Totality Starts at 2:41 / Lasts 2:36

Plan to be at your eclipse viewing site at least two hours before totality starts.

107 Ball Park Road Recreation Center
432 Ball Park Rd, Lexington, SC 29072
33.96928 -81.28138
Totality Starts at 2:41 / Lasts 2:36

108 Gibson Road Soccer Complex
298 29072, 104 Duffie Dr
Lexington, SC 29072
33.97855 -81.25649
Totality Starts at 2:41 / Lasts 2:36

109 Saluda Shoals Park
5605 Bush River Rd
Columbia, SC 29212
34.04793 -81.18824
Totality Starts at 2:41 / Lasts 2:30

110 Frankie's of Columbia
140 Parkridge Dr
Columbia, SC 29212
34.08285 -81.15657
Totality Starts at 2:41 / Lasts 2:27

111 Seven Oaks Park
200 Leisure Ln, Columbia, SC 29210
34.04995 -81.15280
Totality Starts at 2:41 / Lasts 2:29

112 St. Andrews Park
920 Beatty Rd, Columbia, SC 29210
34.05466 -81.11509
Totality Starts at 2:41 / Lasts 2:28

113 Columbia International University
7435 Monticello Rd
Columbia, SC 29203
34.07660 -81.07362
Totality Starts at 2:41 / Lasts 2:24

114 Meadowlake Park
524 Beckman Rd, Columbia, SC 29203
34.07922 -80.99450
Totality Starts at 2:41 / Lasts 2:18

115 Greenview Park and Pool
6700 David St, Columbia, SC 29203
34.07033 -80.99163
Totality Starts at 2:41 / Lasts 2:19

116 Sesquicentennial State Park
9564 Two Notch Rd
Columbia, SC 29223
34.08620 -80.90717
Totality Starts at 2:42 / Lasts 2:12

117 Polo Road Park
730 Polo Rd, Columbia, SC 29223
34.08716 -80.89218
Totality Starts at 2:42 / Lasts 2:10

118 Barnyard Flea Market / RV Park
4414 Augusta Rd, Lexington, SC 29072
33.97639 -81.15858
Totality Starts at 2:41 / Lasts 2:33

119 Columbia Metropolitan Airport
3250 Airport Blvd
West Columbia, SC 29170
33.94252 -81.12107
Totality Starts at 2:41 / Lasts 2:34

120 Riverbanks Botanical Garden
500 Wildlife Pkwy
West Columbia, SC 29169
34.00793 -81.08072
Totality Starts at 2:41 / Lasts 2:29

121 Riverfront Park Amphitheater
Three Rivers Greenway
Columbia, SC 29201
34.00273 -81.05434
Totality Starts at 2:41 / Lasts 2:29

122 Finlay Park
Taylor St, Columbia, SC 29201
34.00689 -81.04090
Totality Starts at 2:41 / Lasts 2:28

123 Olympia Park
1050 Olympia Ave, Columbia, SC 29201
33.98320 -81.03332
Totality Starts at 2:41 / Lasts 2:29

124 Maxcy Gregg Park
Blossom St, Columbia, SC 29201
33.99609 -81.02207
Totality Starts at 2:41 / Lasts 2:28

Plan to be at your eclipse viewing site at least two hours before totality starts.

125 Williams-Brice Stadium
1125 George Rogers Blvd
Columbia, SC 29201
33.97348 -81.01926
Totality Starts at 2:41 / Lasts 2:29

Scan for Map

126 Rosewood Park
901 S Holly St, Columbia, SC 29205
33.98123 -80.99883
Totality Starts at 2:41 / Lasts 2:28

127 Hamilton - Owens Airport
1400 Jim Hamilton Blvd
Columbia, SC 29205
33.97113 -80.99663
Totality Starts at 2:41 / Lasts 2:29

128 Valencia Park
710 S Kilbourne Rd
Columbia, SC 29205
33.98174 -80.98863
Totality Starts at 2:41 / Lasts 2:28

129 Woodland Park
6500 Olde Knight Pkwy
Columbia, SC 29209
33.97205 -80.96417
Totality Starts at 2:42 / Lasts 2:28

130 Hampton Park
1117 Brandon Ave
Columbia, SC 29209
33.98697 -80.95951
Totality Starts at 2:42 / Lasts 2:27

131 Hilton Field
Hartsville Guard Rd
Columbia, SC 29207
34.01679 -80.91239
Totality Starts at 2:42 / Lasts 2:21

132 Bluff Road Park
148 Carswell Dr, Columbia, SC 29209
33.93016 -80.95718
Totality Starts at 2:42 / Lasts 2:30

133 South East Park
951 Hazelwood Rd, Columbia, SC 29209
33.95933 -80.92928
Totality Starts at 2:42 / Lasts 2:28

134 Caughman Road Park
2800 Trotter Rd, Hopkins, SC 29061
33.95855 -80.89499
Totality Starts at 2:42 / Lasts 2:27

Scan for Map

135 Weston Lake Recreation Area
Leesburg Rd, Hopkins, SC 29061
34.00730 -80.83011
Totality Starts at 2:42 / Lasts 2:17

136 Richland County Recreation
2750 McCords Ferry Rd
Eastover, SC 29044
33.94417 -80.68906
Totality Starts at 2:42 / Lasts 2:15

137 Hopkins Park
150 Hopkins Park, Hopkins, SC 29061
33.88485 -80.88877
Totality Starts at 2:42 / Lasts 2:30

138 CP 4545 US-321
Gaston, SC 29053
33.85647 -81.08286
Totality Starts at 2:41 / Lasts 2:36

139 CP I-26
Gaston, SC 29053
33.82188 -81.01485
Totality Starts at 2:42 / Lasts 2:36

140 CP US-21 / Columbia Rd
Gaston, SC 29053
33.79976 -80.97133
Totality Starts at 2:42 / Lasts 2:36

141 Pelion Park
Pelion Park Rd, Pelion, SC 29123
33.72768 -81.26725
Totality Starts at 2:41 / Lasts 2:27

142 CP US-176 / 2688 Old State Rd
Swansea, SC 29160
33.77772 -80.92806
Totality Starts at 2:42 / Lasts 2:36

Plan to be at your eclipse viewing site at least two hours before totality starts.

Scan for Map

143 Longleaf Campground
Congaree National Park
National Park Rd, Hopkins, SC 29061
33.83586 -80.82803
Totality Starts at 2:42 / Lasts 2:31

144 Harry Hampton Visitor Center
100 National Park Rd
Hopkins, SC 29061
33.83012 -80.82268
Totality Starts at 2:42 / Lasts 2:31

145 Poinsett State Park
Lake Poinsett Park Rd
Wedgefield, SC 29168
33.80443 -80.54810
Totality Starts at 2:43 / Lasts 2:24

146 CP US-601 / Colonel Thomson Hwy
St Matthews, SC 29135
33.68336 -80.74312
Totality Starts at 2:42 / Lasts 2:36

147 Orangeburg Dragstrip
194 Dragstrip Rd, Neeses, SC 29107
33.52612 -80.97924
Totality Starts at 2:42 / Lasts 2:19

148 River Oaks Campground
524 Neeses Hwy, Orangeburg, SC 29115
33.48268 -80.90933
Totality Starts at 2:43 / Lasts 2:18

149 Orangeburg County Fairgrounds
350 Magnolia St, Orangeburg, SC 29115
33.48101 -80.85248
Totality Starts at 2:43 / Lasts 2:22

150 South Carolina State University
300 College St NE
Orangeburg, SC 29115
33.49597 -80.85418
Totality Starts at 2:43 / Lasts 2:24

151 Hillcrest Recreational Park
Orangeburg, SC 29118
33.52373 -80.84687
Totality Starts at 2:42 / Lasts 2:28

SE South Carolina

Scan for Map

152 Joe Miller Park
State Rd S-38-972, Elloree, SC 29047
33.53439 -80.57405
Totality Starts at 2:43 / Lasts 2:38

153 Stumphole Landing
18 Halter Ct, Elloree, SC 29047
33.57868 -80.53153
Totality Starts at 2:43 / Lasts 2:35

154 Santee State Park
251 State Park Rd, Santee, SC 29142
33.55202 -80.50183
Totality Starts at 2:43 / Lasts 2:36

155 CP I-95 Lake Marion
Summerton, SC 29148
33.52288 -80.42996
Totality Starts at 2:43 / Lasts 2:35

156 Santee Lakes Campground
1268 Gordon Rd, Summerton, SC 29148
33.51930 -80.42988
Totality Starts at 2:43 / Lasts 2:35

157 Harleyville Community Park
201 S Railroad Ave
Harleyville, SC 29448
33.21082 -80.44898
Totality Starts at 2:44 / Lasts 2:12

158 Gilmore Park
Holly Hill, SC 29059
33.32001 -80.41394
Totality Starts at 2:44 / Lasts 2:29

159 Folk Park
1414 Unity Rd, Holly Hill, SC 29059
33.32885 -80.40038
Totality Starts at 2:44 / Lasts 2:31

Plan to be at your eclipse viewing site at least two hours before totality starts.

Scan for Map

160 Hide Away Campground
Ferguson Landing Way
Eutawville, SC 29048
33.43032 -80.28186
Totality Starts at 2:44 / Lasts 2:35

161 Rocks Pond Campground
108 Campground Rd
Eutawville, SC 29048
33.40307 -80.23254
Totality Starts at 2:44 / Lasts 2:35

162 CP Hwy 45 / 1431 Trojan Rd
Cross, SC 29436
33.37757 -80.14788
Totality Starts at 2:44 / Lasts 2:35

163 Arrow Head
Motel & Campground
1121 Arrowhead Rd, St Stephen, SC
33.40238 -79.86582
Totality Starts at 2:45 / Lasts 2:27

164 Canal WMA Public Field
1123 Arrowhead Rd
St Stephen, SC 29479
33.40430 -79.86381
Totality Starts at 2:45 / Lasts 2:27

165 Andrews Recreation Center
510 W Alder St, Andrews, SC 29510
33.44931 -79.57101
Totality Starts at 2:45 / Lasts 2:00

166 Andrews Park
313 Catclaw Rd, Andrews, SC 29510
33.43681 -79.55267
Totality Starts at 2:45 / Lasts 2:01

167 CP US-52 near Fairsprings Rd
Bonneau, SC 29431
33.28167 -79.96235
Totality Starts at 2:45 / Lasts 2:35

168 Pinopolis Peninsula Park
2421 Pinopolis Rd
Pinopolis, SC 29469
33.24602 -80.03220
Totality Starts at 2:44 / Lasts 2:34

Scan for Map

169 Berkeley County Airport
1003 US-52, Moncks Corner, SC 29461
33.18682 -80.03476
Totality Starts at 2:45 / Lasts 2:33

170 Moncks Corner Rec Complex
418 E Main St
Moncks Corner, SC 29461
33.19498 -80.00172
Totality Starts at 2:45 / Lasts 2:34

171 Old Santee Canal State Park
900 Stony Landing Rd
Moncks Corner, SC 29461
33.19304 -79.97101
Totality Starts at 2:45 / Lasts 2:34

172 Azalea Park
105 W 5th S St, Summerville, SC 29483
33.01492 -80.17963
Totality Starts at 2:45 / Lasts 2:03

173 Gahagan Park
184 W Boundary St
Summerville, SC 29485
33.00832 -80.16465
Totality Starts at 2:45 / Lasts 2:03

174 Charleston Southern University
9200 University Blvd
North Charleston, SC 29406
32.98240 -80.07367
Totality Starts at 2:45 / Lasts 2:06

175 Wannamaker County Park
8888 University Blvd
North Charleston, SC 29406
32.97707 -80.05376
Totality Starts at 2:45 / Lasts 2:07

176 Foster Creek Park
216 Foster Creek Rd
Goose Creek, SC 29445
32.97202 -80.03084
Totality Starts at 2:45 / Lasts 2:08

177 North Woods Community Park
8348 Greenridge Rd
North Charleston, SC 29406
32.95719 -80.05925
Totality Starts at 2:45 / Lasts 2:03

Plan to be at your eclipse viewing site at least two hours before totality starts.

Scan for Map

178 Hillsdale Park
North Charleston, SC 29406
32.93681 -80.04316
Totality Starts at 2:45 / Lasts 2:01

179 Hanahan Softball / Soccer Fields
Bettis Boat Landing Rd
Hanahan, SC 29410
32.93105 -80.02453
Totality Starts at 2:45 / Lasts 2:02

180 MeadWestvaco Park
1206 Bentley Rd
North Charleston, SC 29406
32.90374 -79.98770
Totality Starts at 2:45 / Lasts 2:01

181 Palmetto Islands County Park
444 Needlerush Pkwy
Mt Pleasant, SC 29464
32.86203 -79.83165
Totality Starts at 2:46 / Lasts 2:07

182 Park West Recreation Fields
Park W Blvd, Mt Pleasant, SC 29466
32.88178 -79.78597
Totality Starts at 2:46 / Lasts 2:14

183 Blackbeards Cove
Family Fun Park
3255 N Hwy 17, Mt Pleasant, SC 29466
32.86856 -79.77997
Totality Starts at 2:46 / Lasts 2:12

184 Mount Pleasant KOA
3157 N Hwy 17
Mt Pleasant, SC 29466
32.86478 -79.77596
Totality Starts at 2:46 / Lasts 2:12

185 Buck Hall Recreation Area
Buckhall Landing Rd
McClellanville, SC 29458
33.03922 -79.56159
Totality Starts at 2:46 / Lasts 2:34

186 CP 8931 US-17
McClellanville, SC 29458
33.06112 -79.53816
Totality Starts at 2:46 / Lasts 2:34

Scan for Map

187 Hampton Plantation Historic Site
1950 Rutledge Rd
McClellanville, SC 29458
33.20056 -79.43470
Totality Starts at 2:46 / Lasts 2:25

188 Georgetown Airport
129 Airport Rd
Georgetown, SC 29442
33.31471 -79.32060
Totality Starts at 2:46 / Lasts 2:01

Plan to be at your eclipse viewing site at least two hours before totality starts.

CPSIA information can be obtained
at www.ICGtesting.com
Printed in the USA
LVOW04s1155120217
524027LV00009B/245/P